SpringerBriefs in Energy

SpringerBriefs in Energy presents concise summaries of cutting-edge research and practical applications in all aspects of Energy. Featuring compact volumes of 50 to 125 pages, the series covers a range of content from professional to academic. Typical topics might include:

- A snapshot of a hot or emerging topic
- A contextual literature review
- A timely report of state-of-the art analytical techniques
- An in-depth case study
- A presentation of core concepts that students must understand in order to make independent contributions.

Briefs allow authors to present their ideas and readers to absorb them with minimal time investment.

Briefs will be published as part of Springer's eBook collection, with millions of users worldwide. In addition, Briefs will be available for individual print and electronic purchase. Briefs are characterized by fast, global electronic dissemination, standard publishing contracts, easy-to-use manuscript preparation and formatting guidelines, and expedited production schedules. We aim for publication 8–12 weeks after acceptance.

Both solicited and unsolicited manuscripts are considered for publication in this series. Briefs can also arise from the scale up of a planned chapter. Instead of simply contributing to an edited volume, the author gets an authored book with the space necessary to provide more data, fundamentals and background on the subject, methodology, future outlook, etc.

SpringerBriefs in Energy contains a distinct subseries focusing on Energy Analysis and edited by Charles Hall, State University of New York. Books for this subseries will emphasize quantitative accounting of energy use and availability, including the potential and limitations of new technologies in terms of energy returned on energy invested.

More information about this series at http://www.springer.com/series/8903

Nieves Fernandez-Anez · Blanca Castells Somoza ·
Isabel Amez Arenillas · Javier Garcia-Torrent

Explosion Risk of Solid Biofuels

 Springer

Nieves Fernandez-Anez
Department of Safety, Chemistry
and Biomedical laboratory sciences
Western Norway University
of Applied Sciences
Haugesund, Norway

Isabel Amez Arenillas
Department of Energy and Fuels &
Laboratorio Oficial Madariaga
Universidad Politécnica de Madrid
Madrid, Spain

Blanca Castells Somoza
Department of Energy and Fuels &
Laboratorio Oficial Madariaga
Universidad Politécnica de Madrid
Madrid, Spain

Javier Garcia-Torrent
Department of Energy and Fuels &
Laboratorio Oficial Madariaga
Universidad Politécnica de Madrid
Madrid, Spain

ISSN 2191-5520   ISSN 2191-5539   (electronic)
SpringerBriefs in Energy
ISBN 978-3-030-43932-3   ISBN 978-3-030-43933-0   (eBook)
https://doi.org/10.1007/978-3-030-43933-0

This Springer imprint is published by the registered company Springer Nature Switzerland AG
The registered company address is: Gewerbestrasse 11, 6330 Cham, Switzerland

Acknowledgements

The authors would like to thank everybody that made this research possible. We thank Laboratorio Official J. M. Madariaga and Universidad Politecnica de Madrid for providing the equipment and facilities that have provided the results detailed in this book. Special thanks to the people that helped us during our lab time: Gonzalo Alvarez de Diego, Emilio Garcia Gonzalez and Mariano Perez Calleja.

We thank our friends and family that have supported us during this process. This book would not exist without you.

Part of the research detailed in this book has been produced during Dr. Nieves Fernandez-Anez Ph.D. thesis entitled "Flammability properties of solid biofuels" defended on 15th of April 2016 at Universidad Politecnica de Madrid.

Contents

Part I
Combustible Solids

Chapter 1
Solid Fuels: Fossil and Renewable Combustible Products

Fossil fuel is defined as *"a natural fuel such as coal or gas formed in the geological past from the remains of living organisms"* [1]. In this chapter we are going to focus on its solid form, coal, *"a black or dark brown rock consisting chiefly of carbonized plant matter, found mainly in underground seams and used as fuel"* [2].

In this chapter, a brief description of the origin of fossil fuels, their classification and current use is detailed.

1.1 Origin of Fuel Natural Products: From Peat to Anthracite

The formation of coal was the subject of many discussions at the beginning of the 20th century. However, in 1911 Stevenson [3] already spoke about the huge consensus on the opinions about the origin of coal. He wrote that *"geologists and chemists, with rare exceptions, have recognised that the several types consist mainly of vegetable matter which has undergone chemical exchange"* [3]. He speaks about two theories explaining the formation of coal beds. The doctrine of allochthonous origin, very common but that went into disfavour early in the 19th century, conceives that coal beds are composed of transported vegetable matter deposited in the sea or in lake basins. However, the consensus was around the autochtonous doctrine, which stablishes that the plants which yielded the vegetable matter grew where the coal is found, establishing also the essential need of water for coal formation [4].

Nowadays, the formation of coal is a well-studied process, taught in schools and known for taking millions of years, beginning in areas of swampy wetlands where groundwater is near or slightly above the topsoil. There are two main phases in coal formation: peatification and coalification.

The process of formation of coal is well-known, and started millions of years ago in forestry areas (Fig. 1.1). When the plants in these areas died, they fell into swamp waters, process that continuously took place until there was a thick layer

© The Author(s), under exclusive license to Springer Nature Switzerland AG 2020   3
N. Fernandez-Anez et al., *Explosion Risk of Solid Biofuels*,
SpringerBriefs in Energy, https://doi.org/10.1007/978-3-030-43933-0_1

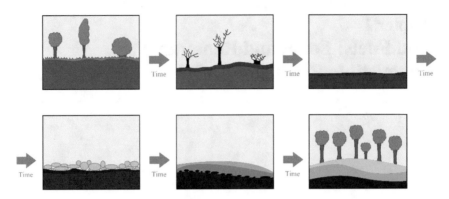

Fig. 1.1 Schematic of the formation of coal

of dead plants rotting in the swamps. This process took place during millions of years, adjusting to the changes of the surface of the Earth, causing layers of different dead material one on top of the other. This partially decayed plant matter is known as peat, and the process dealing to its formation is defined as peatification, mainly dominated by bacterial activity. The weight of the layers packed down the lower layers, generating an increase on heat and pressure that produced chemical and physical changes forcing oxygen out and leaving rich carbon deposits, process known as coalification.

1.2   Types and Characteristics of Coal

Classification of coal can be done depending on their depositional environmental and on the botanical constituents that form coal deposits, existing two different types [5].

Humic coals are developed from largely wood and/or reed/sedge remains through the process of peatification, primarily in the relatively moist aerobic zone.

Sapropelic coals are developed from a largely non-woody source through decay in stagnant or standing water, primarily in the anaerobic zone.

However, the most known categorization of coals is the one depending on their rank: degree of metamorphism or progressive alteration. Coal rank is generally considered to be a function of some combination of heat, pressure and time, related by the Hilt's law. There are four types of coal as shown in Table 1.1.

Lignite is the first product of coalification. The utilization of lignite is difficult mainly because of its high water content, which difficults its storage and transportation. It tends to disintegrate during combustion. However, it is abundant and accessible. It is commonly used for electric power generation near by the mines to reduce the transportation costs.

Table 1.1 Types of coal

	Moisture	Volatiles	Carbon	Darkness	Hardness	Density	Calorific value
Lignite	+ ↑	+ ↑	− ↓	− ↓	− ↓	− ↓	− ↓
Subbituminous							
Bituminous							
Anthracite	−	−	+	+	+	+	+

Table 1.2 Types of coal according to their calorific value [6]

Type of coal	Calorific value (MJ/kg)
Lignite	5.5–14.3
Subbituminous	8.3–25
Bituminous	18.8–29.3
Anthracite	30

Subbituminous. 30% of the coal resources are subbituminous coals. It is used in generating steam for the production of electricity, and thus frequently used in power plants. It can also be liquefied and converted into petroleum and gas.

Bituminous is the most abundant form of coal, dark brown to black in colour and commonly layered. It has a relatively high heat value and low moisture content. It has long been used for steam generation in electric power plants and industrial boiler plants, as well as for making metallurgical coke.

Anthracite is the most highly metamorphosed form of coal. It is black to steel grey and have a brilliant lustre. It is very limited and historically used for domestic heating.

Since the main use of coal is for energy purposes, it can be categorized depending on its calorific value according to the Standard ASTM D388-19a [6], as shown in Table 1.2.

1.3 Use of Coal

The first evidence of the use of coal as a fuel has been found in various ancient sites in China and Wales from 3000 BC. The Roman already used coal for heating and metalworking, but its main achievement is that coal was the fuel that launched the Industrial Revolution. Thus, since the 18th century coal has had a fundamental role in the energy cycle worldwide [7].

Fossil fuels are currently the world's primary energy source, being at the same time the biggest contributor to greenhouse gases emissions. Therefore, governments around the world are engaged in efforts to decrease these emissions by reducing the use of fossil fuels. Partially due to these efforts, world coal production declined in

2014 for the first time this century. This trend continued through 2015 and accelerated in 2016. However, it changed in 2017 with an increase of 3.1%. With the same trend, total global coal consumption in energy terms increased by 1.0% [8].

References

1. Fossil fuel. Oxford dictionary
2. Coal. Oxford dictionary
3. Stevenson JJ (1911) The formation of coal beds. I. An historical summary of opinion from 1700 to the present time. Proc Am Philos Soc 50(198):1–116
4. Jeffrey EC (1915) The mode of origin of coal. J Geol 23(3):218–230
5. O'Keefe JMK et al (2013) On the fundamental difference between coal rank and coal type. Int J Coal Geol 118:58–87
6. ASTM International (2019) ASTM D388-19a. Standard classification of coals by rank. West Conshohocken, PA
7. International Energy Agency (2018) Coal information: overview (2018 edn)
8. Cleveland CJ, Morris C (2014) Section 5. Coal. In: Handbook of energy. Chronologies, tope ten lists, and word clouds, vol II. Elsevier Inc.

Chapter 2
Solid Biofuels

The use of biofuels is subject to great controversy. On one side, environmental policies point biofuels as the key actor of new green technologies to decrease the harmful emissions related with the use of fossil fuels. On the other side, biofuels are accused of competing with food production for land. The reality is that the use of biofuels is rapidly growing around the world, as well as the policies and initiatives directed to their increase.

In this chapter, we are going to focus only on solid biofuels or biomass, defined by the United Nations as: "*non-fossilized and biodegradable organic material originating from plants, animals and micro-organisms. This shall also include products, by-products, residues and waste from agriculture, forestry and related industries as well as the non-fossilized and biodegradable organic fractions of industrial and municipal wastes*" [1].

2.1 Classification

Different classifications for the types of biomass are used worldwide based on their origin or based on their uses and applications. One of the most used classification is the oone established in the Standard EN 14961 [2], withdrawn by Standard ISO 17225 [3], which classified the solid biofuels originated from:

(a) Products from agriculture and forestry
(b) Vegetable waste from agriculture and forestry
(c) Vegetable waste from the food processing industry
(d) Wood waste
(e) Fibrous vegetable waste
(f) Cork waste.

© The Author(s), under exclusive license to Springer Nature Switzerland AG 2020
N. Fernandez-Anez et al., *Explosion Risk of Solid Biofuels*,
SpringerBriefs in Energy, https://doi.org/10.1007/978-3-030-43933-0_2

This Standard [3] classifies biomass in four different groups:

I.   Woody biomass
II.   Herbaceous biomass
III. Fruit biomass
IV.   Blends and mixtures.

However, the most used classification is the one that divides biomass in five categories [4]:

– Wood and woody biomass: *"by-product of management, restoration, and hazardous fuel reduction treatments, as well as the product of natural disasters, including trees and woody plants"* [5].
– Herbaceous biomass: *"from plants that have a non-woody stem and which die back at the end of the growing season. It includes grains or seeds crops from food processing industry and their by-products such as cereal straw"* [6].
– Aquatic biomass: is composed of diverse species of micro- and macroalgae and aquatic plants. They have been suggested as good candidates for the production of fuels because of their higher photosynthetic efficiency, higher biomass production, and faster growth compared to lignocellulosic biomass types [7].
– Animal and human waste biomass: wastes from both animal and humans can be treated and reused as biomass, for example treating sewage sludge from wastewater treatment plants.
– Biomass mixtures: any mixture of two or more different types of biomass. These mixtures have properties that do not have to be the sum of the individual biomass, but can be improved or worse.

2.2   Composition and Characteristics of the Different Types of Biomass

Biomass essentially consists of macromolecular organic polymers. It is composed by three major components, with a small amount of extractives and ash.

- Cellulose is the most common organic substance. It is a polysaccharide consisting of glucose chains which are held together by hydrogen bonds in crystalline clusters, linked by glycosidic linkage, forming the framework of the cell walls. Those hydrogen bonds between hydroxyl groups, cellulose is not soluble in water. Cellulose is composed only by β-glucose monomers whose structure can be written as $(C_6H_{10}O_5)_n$. Regarding biomass, approximately half of its composition is cellulose (Fig. 2.1).
- Hemicellulose is structurally similar to cellulose, but also contains other sugar types as basic building blocks such as xylose, galactose, arabinose, mannose and rhamnose. Hemicellulose structure is simpler than cellulose, and it is an amorphous polymer. Cellulose and hemicellulose serve as structural components

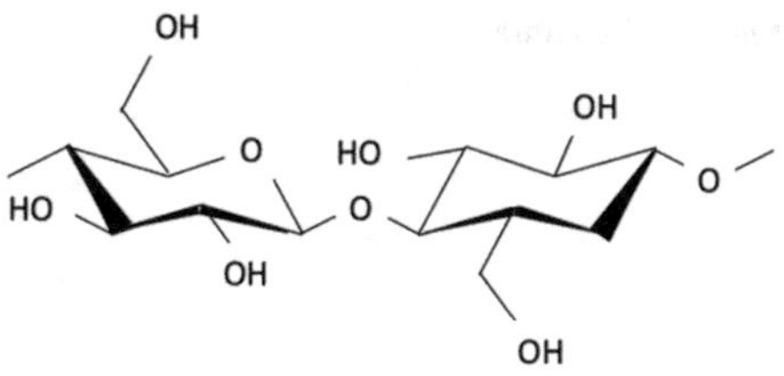

Fig. 2.1 Structural formula
for cellulose

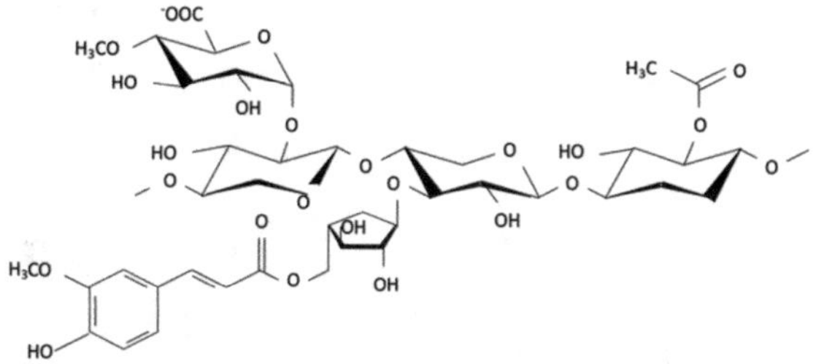

Fig. 2.2 Structural formula
for hemicellulose

of the plant cell wall. As hemicellulose is composed by different polysaccharides
that change from different plants, it is difficult to write its chemical structure
however, according to Hongzhang, it can be written as $(C_5H_8O_4)_n$ [8] (Fig. 2.2).

- Lignin is a three-dimensional aromatic branched-chain macromolecule that acts
  as a binder for the cellulosic tissue as it binds the cell fibres and vessels providing
  structural support [9]. Lignin is composed by substituted phenyl propane units: it
  is a complex phenylpropanoid polymer. The monomeric precursors that synthe-
  sized lignin are p-coummaryl alcohol, coniferyl and sinapyl alcohol. When those
  precursors are incorporated into the polymers that compose lignin they become
  monomeric units called p-hydroxyphenyl, guaiacyl, and syringyl respectively, and
  its chemical composition can be written as $(C_{31}H_{34}O_{11})_n$ [11] (Fig. 2.3).

The contents of cellulose, hemicellulose and lignin in biomass vary significantly
depending on the type of biomass. The cellulose content varies between 40–60%, the
hemicellulose content is 15–30%, and the lignin content is approximately 10–25%
[11].

When biomass started to become important, and their parameters and characteristics started to be studied and determined, it was thought that the methodology and
logic from coal experiments could be applied to biomass [12]. However, once experiments started, and different types of biomass started to apply and be more common,
this statement proved to be wrong. Additional problems also occur in many biomass
investigations due to the use of unsuitable scientific approaches, incomplete data or
unusual and sometimes inappropriate terms that lead to inaccurate interpretations
and misunderstandings about the biomass and biomass fuels.

As a comparison between a generic biomass and coal fuels, Demirbas [13] showed
the main differences between physical, chemical and fuel properties of biomass and
coal fuels as shown in Table 2.1.

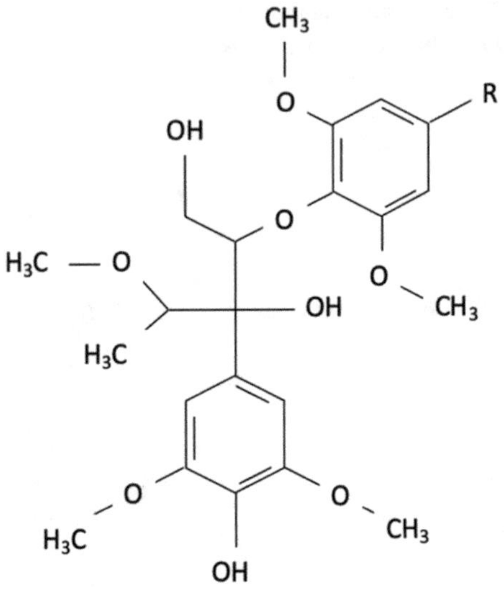

Fig. 2.3 Structural formula
for lignin

Table 2.1 Differences on
physical, chemical and fuel
properties of biomass and
coal

Property	Biomass	Coal
Fuel density (kg/m^3)	~500	~1300
Particle size	~3 mm	~100 μm
C content (wt% of dry fuel)	42–54	65–85
O content (wt% of dry fuel)	35–45	2–15
S contents (wt% of dry fuel)	Max 0.5	0.5–7.5
SiO$_2$ content (wt% of dry fuel)	23–49	40–60
K$_2$O content (wt% of dry fuel)	4–48	2–6
Al$_2$O$_3$ content (wt% of dry fuel)	2.4–9.5	15–25
Fe$_2$O$_3$ content (wt% of dry fuel)	1.5–8.5	8–18
Ignition temperature (K)	418–426	490–595
Peak temperature (K)	560–575	–
Friability	Low	High
Dry heating value (MJ/kg)	14–21	23–28

A common way of representing the composition of organic materials is a triangular graph, where the three vertices of a triangle represent volatiles (top), ash (left) and fixed carbon (right) contents. In Fig. 2.4 some samples are represented: biomass samples are located nearby the top vertex, with high volatiles content and low ash content. Waste (sewage sludge in this case) are located nearby the right side of the triangle due to its much larger ash content and relatively large fixed carbon content. Finally, coal is represented in a larger area of the diagram, depending on the rank. Low rank coals have lower fixed carbon content and are located nearby the centre of the triangle. When increasing the rank, fixed carbon content increases, and the

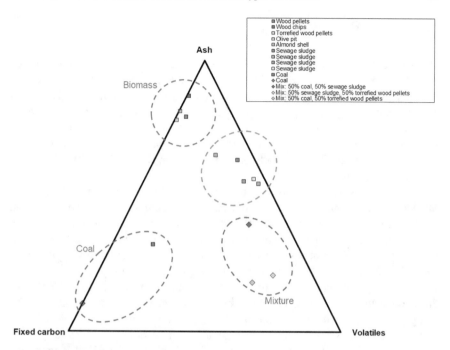

Fig. 2.4 Triangular diagram for biomass, waste and fixed carbon [14]

samples are located nearer the left side of the graph, as the coke sample observed in Fig. 2.4.

Looking at the elements forming the biomass, the chemical composition can be simplified as a combination of the major elements (content higher than 1.0%) that are C, O, H, N, Ca and K, minor elements (content between 0.1 and 1.0%) that normally include Si, Mg, Al, S, Fe, P, Cl and Na, and several trace elements (content lower than 0.1%) that vary from origin to origin [15].

Obernberger et al. [16] already reported the similarities in concentrations of C, H and O among the most common Europena biofuels (wood, bark, straw and cereals). However, the inorganic elements such as the main ash forming constituents (Si, Ca, Mg, K, Na, P, S, Cl, Al, Fe, Mn) or heavy metals (Cu, Zn, Co, Mo, As, Ni, Cr, Pb, Cd, V, Hg) show significant differences.

Vassilev et al. [15] reviewed the composition of 86 varieties of biomass (148 samples). It is important to notice here that the mentioned review does not include a high number of samples, since the amount of data available in literature is not large. They also concluded on the similar contents of C, H and O and the significant differences in contents of ash-forming elements. However, these differences vary depending on the studied biomass.

2.3 Use of Biomass

The World Bioenergy Association [17] with statistics from the International Energy Agency (IEA) pointed out in their last annual report that, despite the efforts made, during 2016–2017 the supply of fossil fuels has increased more than the supply of renewable energy. The percentage of total primary energy supply covered by renewable energy worldwide was 13.9% in 2017, only 0.9% higher than the one for 2000. Looking at the gross final energy consumption, the percentage covered by renewable energy in 2017 was only 0.3% higher than the one in 2000 (17.9% and 17.6% respectively).

These numbers are not increasing at the level that is expected with all the policies that have appeared during the past two decades, some of them that will be discussed in part II of this book. However, focusing on the specifics of the renewable energies, biomass (solid, liquid and gaseous) has a major share in renewables. Worldwide, consumption of biomass has increased in more than 10 EJ in these past 17 years, even if its contribution has been significantly reduced, from representing 78.25% of the total primary energy supply from renewable energies in 2010 to 68.58% in 2017. Nevertheless, biomass has 96% share in the renewable heat market globally, being almost half of all energy consumed in the form of heat.

Looking at the situation in the European Union according to Eurostat [18], it is estimated that two third of current energy production in the EU28 is produced by biomass sources, solid, gaseous and liquid. The primary production of renewable energy within the EU28 increased by 64.0% between 2007 and 2017. Among renewable energies, the most important source was wood and other solid biofuels, accounting for 42.0% of primary renewables production in 2017.

References

1. United Nations Framework Conventions on Climate Change. Annex 8. Clarifications on definition of biomass and consideration of changes in carbon pools due to a CDM project activity
2. EN 14961-1 : 2010. Solid biofuels—fuel specifications and classes. Part 1—general requirements
3. ISO 17225-1:2014 (2014) Solid biofuels—fuel specifications and classes. Part 1—general requirements
4. Tursi A (2019) A review on biomass: importance, chemistry, classification, and conversion 22:962–979
5. Forests and Rangelands (2019) Woody biomass utilization and the WBUG. https://www.forestsandrangelands.gov/woody-biomass/overview.shtml. Accessed 30 Dec 2019
6. IPCC special report on renewable energy sources and climate change mitigation summary for policymakers (2011) pp 5–8
7. de Wild PJ (2015) Biomass pyrolysis for hybrid biorefineries. In Industrial biorefineries & white biotechnology. Elsevier, pp 341–368
8. Chen H (2015) Lignocellulose biorefinery engineering. Woodhead Publishing Limited
9. Spliethoff H (2010) Power generation from solid fuels. Springer

10. Cleri F (2016) Undergraduate lecture notes in physics, the physics of living systems. Undergraduate Lecture Notes in Physics (ULNP)
11. Want S, Dai G, Yang H, Luo Z (2017) Lignocellulosic biomass pyrolysis mechanism: a state-of-the-art review. Program Energy Combust Sci 62:33–86
12. Baxter LL (1993) Ash deposition during biomass and coal combustion: a mechanistic approach. Biomass Bioenergy 4(2):85–102
13. Demirbas A (2004) Combustion characteristics of different biomass fuels. Program Energy Combust Sci 30(2):219–230
14. Warnsloh JM (2015) TriAngle: a Microsoft excel^{TM} spreadsheet template for the generation of triangular plots. Neues Jahrb für Miner J Miner Geochem 192(1):101–105
15. Vassilev SV, Baxter D, Andersen LK, Vassileva CG (2010) An overview of the chemical composition of biomass. Fuel 89(5):913–933
16. Obernberger I, Biedermann F, Widmann W, Riedl R (1997) Concentrations of inorganic elements in biomass fuels and recovery in the different ash fractions. Biomass Bioenergy 12(3):211–224
17. Lang A, Bradley D, Gauthier G (2016) Global bioenergy statistics, world bioenergy association
18. E Commission. Eurostat. http://ec.europa.eu/eurostat/web/energy/statistics-illustrated

Chapter 3
Risk in Handling Solid Fuels

Regardless the goal and final use of solid fuels, all these substances need to be handled, processed and stored. These steps involve several risks that need to be identified, characterised and treated in order to avoid, or at least reduce, the consequences of a possible accident. In this chapter, we are going to focus in two main consequences that can derive from the treatment and storage of both fossil fuels and biofuels: self-ignition and explosions.

3.1 Self-ignition

During the handling and storage of solid substances, the oxygen of the air is continuously interacting with these substances through adsorption, causing their oxidation. This process, which produces a heat source as it is exothermic, is known as self-heating, since the phenomenon occurs without any external influence as a spark or a flame. Part of this heat accumulates inside the material, accelerating the oxidation reaction. If the heat generated is higher than the heat that can be dissipated into the surrounding environment, the temperature of the sample increases. This process feeds itself, and it can cause the combustion of the material. This is called self-ignition or spontaneous combustion.

As a definition to be used from hear, self-ignition or spontaneous ignition is the ignition of a material without any external spark, flame or any other ignition source.

3.1.1 The Phenomena of Self-ignition and Spontaneous Combustion

The phenomenon of ignition has been studied since the end of the XIX Century, being the topic of some of the earliest-recorded scientific studies. However, quantitative studies started much later. The nature of the self-heating problem is that there are

© The Author(s), under exclusive license to Springer Nature Switzerland AG 2020
N. Fernandez-Anez et al., *Explosion Risk of Solid Biofuels*,
SpringerBriefs in Energy, https://doi.org/10.1007/978-3-030-43933-0_3

basically two practical questions: (i) will this material ignite, and if the answer is yes, (ii) how long will this take. As Babrauskas pointed out in 2007 [1], and we believe is still a correct statement nowadays, there are no reliable methods for answering the second question for engineering purposes. A number of theoretical treatments have been suggested but their results differ by several orders of magnitude and none has emerged as being the correct one.

The main work on self-heating, self-ignition and spontaneous combustion has been historically done for coals, since this phenomenon is, and has been, the starting point of a large number of accidents occurred in mines.

One of the first scientists describing the steps that a spontaneous combustion process goes through was van Krevelen [2], who, in 1961, defined a four-step process:

- Chemisorption of oxygen with increasing weight—up to 70 °C.
- Initial release of oxidation reactions products and inner water −70 to 150 °C.
- Production of larger amounts of oxidation reaction products −150 to 230 °C.
- Fast burning including production of soot—above 230 °C.

Later, in 1975, Marinov [3] made an attempt to elucidate the mechanism of coal self-ignition at low temperatures, and observed that:

- The explosion near 200°C is due to the dissociation of ether bonds.
- Exothermic effect in the range of 400–450°C is due to the break-down of methyl groups bound to aromatic rings.
- Anthracite self-ignition at 440°C is related to the dissociation of aromatic C–H bonds.

In 1984, Bowes [4] studied the heat transfer occurred during self-heating and defined an equilibrium curve depending on the storage size, where 4 regions can be observed, as shown in Fig. 3.1 and in Table 3.1, where self-heating and self-

Fig. 3.1 Equilibrium states in a porous solid presenting a self-heating process due to atmospheric oxidation at room temperature. Adapted from [5]

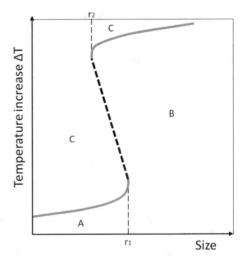

Table 3.1 Different behaviours of storage of solid fuels

		Increase temperature			
		A	B	C	D
Increase size	$r_1 < r$		Self-ignition		Self-ignition
	$r_2 < r < r_1$	Self-heating	Self-ignition	Self-heating	Self-ignition
	$r < r_2$	Self-heating		Self-heating	

ignition can occur under many circumstances, since materials oxidize even at very slow speeds.

It is thought that self-ignition might account for around 12% of the fires that occur in underground mines, and 6% of those that occur in the agri-food industry [6]. Self-ignition not only involve fires, but also explosions, which can occur when the dust originated in the handling of materials results in the formation of an explosive atmosphere as it is explained at the end of this chapter.

3.1.2 Mathematical Models Explaining the Phenomena

Three mathematical models are used to analyse oxidation and self-heating reactions, all of them assuming a single step global reaction, constant thermal properties of the material, no reactant consumption and no restriction of oxidizer availability (Fig. 3.2).

Semenov's model assumes a one-step Arrhenius temperature dependence with uniform temperature distributions, neglecting consumption of the reactant material.

Frank-Kamenetskii does not assume uniform temperature profile. The theory incorporated the heat conduction through the material due to the chemical release to Semenov's model.

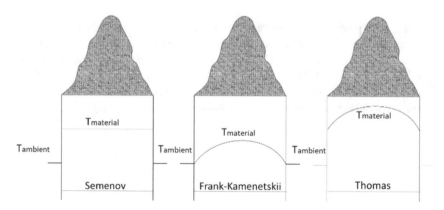

Fig. 3.2 Temperature distributions of different mathematical models for self-heating processes. Adapted from [5]

Finally, Thomas's model builds upon Frank-Kamenetskii model by additionally considering the convective heat losses effects from the surface.

Here, we are going to focus on Frank-Kamenetskii's theory, which is the most used one since the basket's experimental methodology is based on this model, and the theory can be used to extrapolate small-scale experiment to large-scale situations.

This theory assumes that the material studied is reactive and one dimensional. To solve the transient heat conduction equation, Frank-Kamenetskii theory establishes the following equation, expressing the reaction rate as the Arrhenius law for dependence on temperature. With this equation, the dependence of critical sample size and ambient temperature that is the base of the basket experiments is obtained.

$$ln\left[\frac{\delta_c \cdot T_{a,c}^2}{r^2}\right] = Ln\left[\frac{Q \cdot E \cdot f}{R \cdot \kappa}\right] - \frac{E}{R \cdot T_{a,c}}$$

where δ_c is the critical value of the dimensionless parameter determined by the following equation, $T_{a,c}$ is the critical ambient temperature for which self-ignition occurs, E is the activation energy, κ is the effective thermal conductivity of the sample, R the universal gas constant, Q is the heat of reaction per fuel mass, and f is the value of the mass action law which relates the concentration of fuel and oxygen at the initial time to reaction rates.

$$\delta = \frac{QEr^2 f(c_o)e^{-E/RT_a}}{\kappa RT_a^2}$$

3.1.3 Factors Influencing the Self-ignition and Spontaneous Combustion of Solid Fuels

Predicting the propensity of coal to self-heat is complex and difficult due to the number and variety of variables involved. The main factors that influence this risk can be divided into intrinsic, the ones that are internal properties of the materials, and extrinsic ones, not dependent on the fuels, and are detailed in Table 3.2 (This report: https://www.usea.org/sites/default/files/102010_Propensity%20of%20coal%20to%20self-heat_ccc172.pdf).

	Intrinsic factors	Extrinsic factors
Table 3.2 Factors influencing self-heating of coals	Volatile matter	Segregation/accumulation of fines
	Moisture	Maintenance
	Friability	Oxygen access
	Pyrite	Degree of consolidation
	Particle size	Exposed stack surface
	Micro-fracturing	Relative moisture content
	Geological disturbances	Reduction of oxygen concentration through emission of gases

3.1.4 Empirical Determination of Self-ignition Tendency

The procedures that we review below are the ones that study the self-heating of solids by a homogeneous increase on the temperature of the ambient air surrounding the sample. These procedures ensure that no ignition source, as flame or spark, cause or promote the ignition of the samples, and the only variable that affects is the increase on the surrounding temperature. There are three main procedures to determine the self-ignition tendency of a substance: isothermal calorimetry, isothermal oven procedures and thermogravimetric analyses. Additionally, off-gassing most probably occurs together with self-heating, so the self-heating tendency might be estimated early detection of off-gassing during storage [23]. The decomposition reaction of an organic material can be assumed to be first order reactions. The overall reaction can be written as:

$$\text{Sample} + O_2 \rightarrow CO_2 + O_2 + CH_4 + q$$

where q is the heat generated during the process. This reaction assumes complete combustion, and CO production should be added to the emitted gases when combustion is incomplete.

Isothermal calorimetry is a physical technique used to determine the thermodynamic parameters of substances. In isothermal calorimetry the sample material is contained at a fixed temperature and the produced heat is conducted away for quantification, most isothermal calorimeters measure thermal power. Heat flow from the sample trough a thermal resistance to a heat sink establishes a temperature difference across the thermal resistance, which after calibration provides a measure of the rate of heat generation. Rates of heat generation that can present a self-heating hazard are, however, very low and micro calorimeters capable of measuring thermal powers of order 10^{-6} W or less are required [4].

For the isothermal oven procedure, the material is introduced into a cube shaped basket from 50 to 150 mm side made of stainless steel mesh. The basket is suspended in an air circulating oven, preheated to a known temperature. Temperature is measured both inside and outside the basked and plotted against time during the

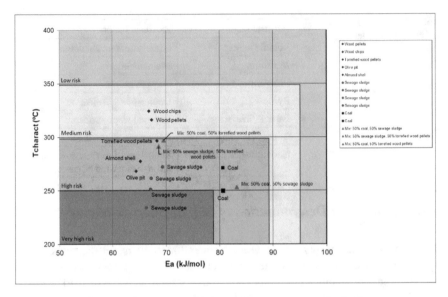

Fig. 3.3 Thermal susceptibility graph: oxidation temperature versus activation energy

test., with thermocouples both inside out outside the basket to record temperature. Two types of behaviour are observed: sub-critical, where the sample temperature increases above oven temperature, but then slowly drops back after reaching a peak; super-critical, where the sample temperature continues to rise above oven temperature and heats to ignition. By plotting a graph of $\ln(2.52 * T^2/r^2)$ versus $1/T$, the self-heating characteristics of the material may be determined: apparent activation energy and Arrhenius factors. The results can then be scaled to full size plant through thermal ignition theory or by computer.

The experimental procedure for thermogravimetric analyses is deeper explained in Chap. 4. As a brief introduction, thermogravimetric analyses consist on the measurement of weight of a sample during a controlled heating process. It measures the mass of the sample during this heating process, and the heating process can be defined. With the combination of two parameters obtained through these analyses, the oxidation temperature and the activation energy, a thermal susceptibility graph as the one showed in Fig. 3.3 can be obtained. In this graph, different areas are delimited, showing four different degrees of self-ignition risk.

3.2 Smouldering

Smouldering is one of the most probable subsequence of self-ignition processes. It is defined as the slow, low temperature flameless form of combustion, sustained by the heat evolved when oxygen directly attacks the surface of a condensed-phase fuel [7, 8].

The fundamental difference between smouldering and flaming combustion is that, in the first one, both the oxidation and the consequent heat release occur on the solid surface of the fuel, while in flaming combustion these occur in the gas phase surrounding the fuel [9]. Temperatures, spread rate and heat released on smouldering fires are lower than the observed in flaming fires. Because of its low temperature, smouldering is an incomplete reaction that emits a mixture of toxic, asphyxiant and irritant gases and particulates at a higher yield than flaming fires.

Smouldering combustion is the most difficult type of combustion to extinguish, requiring much larger amounts of water than flaming fires [10], lower oxygen concentration and longer holding times when suppressing

Smouldering and flaming are closely related, and one can lead to the other. The transition from smouldering to flaming is a spontaneous gas-phase ignition supported by the smoulder reaction which acts both as the source of the gaseous fuel and of heat to carry the reaction.

Smouldering of stored organic samples is a common phenomenon in silos or any other big storage where these samples are left during long times. The detection of this type of combustion is difficult due to their characteristics previously mentioned. Low emission of smoke, low temperatures and slow processes make them almost impossible to detect in many situations, but they can both consume the whole sample stored, as well as cause worse consequences as a dust explosion.

3.3 Dust Explosions

Self-ignition not only involves fire, but also explosions, which can occur when the dust originated in the handling of materials results in the formation of an explosive atmosphere.

Alongside vapour cloud explosions (VCE) and boiling liquid expanding vapour explosions (BLEVE), dust explosions pose the most serious and widespread of explosion hazards in the process industries [14]. There are seven key differences between combustible dusts and flammable gases: (i) necessary conditions for a deflagration, (ii) chemical purity of fuel, (iii) particle size and shape, (iv) uniformity of fuel concentration and initial turbulence, (v) range of ignitable fuel concentrations, (vi) heterogeneous and homogeneous chemical reactions, and (vii) incomplete combustion.

An explosive atmosphere is defined as a mixture between air and fuel substance, in which the ignition spreads to the entire mixture producing an explosion. This propagation may be a deflagration, subsonic combustion with pressure between 1 and 10 bar, or a detonation, supersonic combustion with pressure above 10^5 bar. Regarding dust clouds or hybrid mixtures (dust and gases), as happens with solid biofuels, deflagration is the typical explosion propagation form. The propagation of the combustion across the dust cloud can occur through oxidation of the particle surface itself, or by combusting the gases emitted by the particles when heated until pyrolysis.

The first documented dust explosion occurred in 1785 in a warehouse in Turin, where the flour used for baking exploded causing two people injured and material losses in the bakery (broken windows…) [11].

In 1845, Faraday demonstrated that coal dust could ignite and be the cause of an explosion without any combustible gas involved in the process [12]. Thanks to this, since 1908 the Bureau of Mines started investigating dust explosions in coalmines.

This shows that dust explosions have been a recognised threat to humans and properties for a long time. We know that almost all combustible materials, when they are fine enough, can evolve into an explosion. However, it is still an unknown threat for many people working with or nearby dusts. There are several "myths" around dust explosions that are widely believed and that need to be clarified in order to ensure safe working spaces [13]. Additionally, the appearance and development of new solid fuels represent also new threats that are still unknown and need to be studied and treated.

The explosion pentagon (Fig. 3.4) affords us everything we need to know on a fundamental level about dust explosion causation.

As the explosion pentagon shows, for a dust explosion to start, a dust cloud needs to be present in a confined space with an ignition source. Not only this, the dust cloud must present a certain state to be the cause of an explosion. This state will always be dynamic, so it is difficult to characterise. This cloud will always be subjected to gravity and other inertia forces, making difficult to determine its state in each situation. However, when the concentration of dust in the cloud is in the explosibility range, and an external ignition source with the needed conditions is present, a primary dust explosion can occur.

Fig. 3.4 Dust explosions pentagon

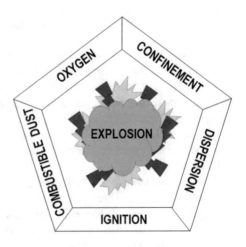

Regarding ignition sources that may trigger a dust explosion, it is necessary to consider smouldering or burning nests, incandescent particles, open flames, hot surfaces, heat (mechanical impacts), sparks (electric, electrostatic discharges, friction, impact), shock waves, static electricity and lightning [14].

Smouldering or burning nest might be produced due to self-heating process, in which the exothermic reactions increase the heat released elevating the temperature. Self-heating is a slow process, but it might be accelerated because of autocatalytic reactions and produce burning nests. Incandescent particles have a similar process: once that it is produced, it continues releasing heat, or even burning for a long period of time.

Open flames are an obvious ignition source and very effective one for cloud ignition. Not so evident are the hot surfaces, which may overheat to ranges between 100–200°C and produce the inflammation of material in contact. Hot surfaces can ignite dust that disperses into a cloud afterwards, or even ignite a dust layer producing a burning nest. Only when the hot surfaces reach 400 °C, the cloud might explode considering the hot surface the main trigger [15].

Regarding the heat released by mechanical impacts, it is quite common in industry, as it can be produced by simple operations such as cutting or welding. Those mechanical works can also produce sparks because of friction or impact. Of course, sparks can also be produced by electricity or electrostatic discharges or arcs.

A shock wave compresses and heats the mixture until the explosion is produced. In those cases, the compression is so considerable the temperature reaches 1000 K because of increasing pressure [16].

Lightning is indeed a static electricity process. A lightning stroke may ignite fuel materials because of the energy released. On the other hand, according to [17] static electricity may produce sparks discharging 0.25 mJ electricity enough to produce the ignition.

A dust explosion is initiated by the rapid combustion of flammable particulates suspended in air. If the ignited dust cloud is unconfined, it would only cause a flash fire. But if it is confined, even partially, the heat of combustion may result in rapid development of pressure, with flame propagation across the dust cloud and the evolution of large quantities of het and reaction products. The furious pace of these events results in an explosion. A dust explosive atmosphere will be characterized by different dust parameters such as particle size distribution, homogeneous concentration, composition, dust cloud turbulence, ignition temperature, etc. [18–20].

Granulometry has an important role in dust explosions. Small particle sizes increase the contact surface area, which means that the combustion rate is increased. Also, it is easier for small particle size to stay suspended in a dust cloud, so to achieve the concentration that can lead to an explosion. The flame front will not be homogeneous as the particles are gross and produces a heterogeneous front. The combustion zone will be wider, and the dominant heat transfer mechanism will be thermal radiation. In some cases, when the particle size is small enough, pyrolysis and devolatilization process (which starts combustion in organic dusts) occur faster and the gas phase combustion is the phenomenon that governs the dust explosion.

Particle size might be the most important parameter in order to characterize bio-fuels dust cloud explosions. As it is been said, compaction, density or particle aggre-gation are defined by the particle size distribution but, besides that, granulometry defines the available surface area to oxidize, so the reaction velocity. It also affects to the dust concentration of the cloud. According to Eckhoff [11], this concentration must be between 100 g/m^3 and 2 kg/m^3 in order to have a mixture appropriate for explosion: beneath the lower limit, fuel will not be enough, and above the upper limit oxygen will not be enough.

Besides granulometry and concentration, substance composition, flame propaga-tion or moisture, are critical factors. Regarding composition, the fuel needs to be oxidizable so the exothermic reaction can occur, and it is fast enough to generate heat before it is dissipated in the surrounding environment. The minimum ignition temperature required to produce the combustion of a substance will depend on the composition of the same. Agricultural and forestry biomass has minimum ignition temperatures between 400 °C and 500 °C [21], and sewage sludge between 420 °C and 510 °C [22]. However, if the other factors (humidity, granulometry, turbulence, etc.) are appropriate enough, the temperature can be diminished to 200 °C. On the other hand, the amount of oxidant is also important. As it happened when defining fire parameters, for explosions a concentration of, at least, 20.5% is required so there is enough oxygen to be consumed during the combustion reaction.

Finally, the role of turbulence is also important as it affects to other factors such as dispersion or concentration. High turbulence will facilitate the dispersion of the dust in the cloud, so when the ignition source produces the inflammation of the cloud, it will propagate faster as the turbulence will mix the burning and the unburnt particles. High turbulence will make no difference between zones: the cloud will be completely heterogeneous. High turbulence is important at the beginning of the explosion, when the ignition is produced. Afterwards, low turbulence will benefit the heat release as it will not dissipate it [21].

References

1. Babrauskas V (2007) Ignition: A century of research and an assessment of our current status. J Fire Prot Eng 17(3):165–183
2. van Krevelen DW (1961) Coal: Typology, chemistry, physics, constitution: with 76 tables. Elsevier
3. Marinov V (1975) Studies on the self-ignition of high-molecular substances at low tempera-tures. Part II. Self-ignition and oxidative degradation of coals and anthracite. J Therm Anal Calorim 7(2):333–345
4. Bowes PC (1984) Self-heating: evaluating and controlling the hazards. Department of the Environment, Building Research Establishment
5. Torrent JG, Madariaga LOJM (2003) Seguridad industrial en atmósferas explosivas. Labora-torio Oficial José María Madariaga
6. García-Torrent J, Ramírez-Gómez Á, Querol-Aragón E, Grima-Olmedo C, Medic-Pejic L (2012) Determination of the risk of self-ignition of coals and biomass materials. J Hazard Mater 213:230–235

7. Ohlemiller TJ (1986) Smoldering combustion. Center for Fire Research
8. Ohlemiller TJ (1985) Modeling of smoldering combustion propagation. Program Energy Combust Sci 11(4):277–310
9. Rein G (2009) Smouldering combustion phenomena in science and technology
10. Hadden R, Rein G (2011) Burning and suppression of smouldering coal fires. In: Coal and peat fires: a global perspective, pp 317–326
11. Eckhoff RK (2005) Explosion hazards in the process industries
12. Palmer KN (1973) Dust explosions and fires [by KN Palmer]
13. Amyotte PR (2014) Some myths and realities about dust explosions. Process Saf Environ Prot 92(4):292–299
14. Abbasi T, Abbasi SA (2007) Dust explosions–cases, causes, consequences, and control. J Hazard Mater 140(1):7–44
15. Eugene Musgrave G, Larsen AM, Sgobba T (2009) Safety design for space systems
16. Glassman I, Yetter RA (2008) Combustion, 4th edn. Elsevier
17. Nolan DP (2011) Handbook of fire and explosion protection engineering principles. Elsevier
18. Ebadat V (2010) Dust explosion hazard assessment. J Loss Prev Process Ind 23(6):907–912
19. Eckhoff RK (2009) Understanding dust explosions. The role of powder science and technology. J Loss Prev Process Ind 22(1):105–116
20. Yuan Z, Khakzad N, Khan F, Amyotte P (2015) Dust explosions: A threat to the process industries. Process Saf Environ Prot 98:57–71
21. Abbasi T, Abbasi SA (2007) Dust explosions-Cases, causes, consequences, and control. J Hazard Mater 140(1–2):7–44
22. Fernandez-Anez N, Garcia-Torrent J, Medic Pejic L (2014) Flammability properties of thermally dried sewage sludge. Fuel 134:636–643
23. Sedlmayer I, Arshadi M, Haslinger W, Hofbauer H, Larsson I, Lönnermark A, Nilsson C, Pollex A, Schmidl C, Stelte W, Wopienka E (2018) Determination of off-gassing and self-heating potential of wood pellets–Method comparison and correlation analysis. Fuel 234:894–903

Chapter 4
Flammability Parameters

Any industrial facility that generates or handles any type of solid biofuels should design prevention and protection measurements based on both the ignition sources that may be present and the characteristics of the dusts. These characteristics highly vary between substances and even among the same substance, so their determination is essential and should be complete at the beginning of their use.

According to the nature of the characteristics, they may be divided in the following groups:

- General characteristics.
- Ignition sensibility.
- Explosion severity.
- Thermal susceptibility.
- Thermal stability.
- Hazardous materials' transport.

4.1 General Characteristics

This group is made by two broadly studied and well-known characteristics: moisture and particle size.

4.1.1 Moisture Content

Moisture presents an antagonistic effect in dust explosions that have made this parameter one of the most studied ones. Depending on the chemical composition of the dust, moisture inhibits or promotes the ignition of the substance and the severity of its potential explosion [1]. In the case of biofuels, we are looking at the first situation. At lower moisture content, the moisture would mainly consume the reaction heat of

© The Author(s), under exclusive license to Springer Nature Switzerland AG 2020
N. Fernandez-Anez et al., *Explosion Risk of Solid Biofuels*,
SpringerBriefs in Energy, https://doi.org/10.1007/978-3-030-43933-0_4

dust explosion by temperature rise and phase change. In this situation, because the heat consumption is proportional to the mass of moisture, the measured explosion severity reduces gently and linearly with the rising moisture content. Nevertheless, as the moisture content continues to rise, due to the stronger interparticle cohesion between particles, besides consuming heat, the existence of moisture would also cause the agglomerations of dust particles and, thereby, increase the effective particle size of dusts and weaken dispersion of dust cloud, so that the reduction of explosion becomes more remarkable and even dust cloud cannot ignite [2].

4.1.2   Particle Size

Looking at the particle size, it is well known that, for coal samples, fines facilitate the flame propagation [3]. The easiness of ignition of a dust cloud and the severity of the explosion increases when the particle size is smaller as the contact surface increases [4]. It has been observed that if the particles in an organic dust are really small and the devolatilisation no longer control the explosion rate, further particle size reduction will not increase the overall combustion rate further. That is why for many metals the limiting particle is considerably smaller than for most organic materials and coals: metals do not produce an homogeneous combustible gas phase by devolatilisation as organic materials do [5] (Fig. 4.1).

This tendency was believed to be truth, and not further questioned for several years. However, when light biomass is stored or deposited in layers, it has shown different results than the expected ones. Looking at the flammability tendency of layers, the behaviour observed for pellets and their correspondent forming dust are

Fig. 4.1 Influence of particle size on the explosibility of dusts. Adapted from [5]

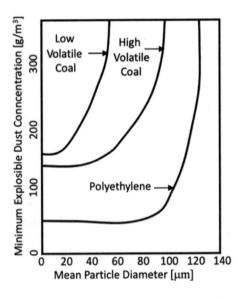

similar, being the same, or even more sensible in the case of pellets in some materials [6]. Moreover, this discrepancy has been also observed when studying the self-heating tendency of biomass. Restuccia et al. [7] observed that wheat biomass, both fine dust and pellets present similar behaviour, being the temperatures observed during this process equal. This one more reflects the need of a deeper, longer and more focus research that builds a complete understanding of the behaviour of solid biofuels in order to ensure safe working environments.

Nevertheless, the influence of particle size is clear in the case of explosion of dusts. When particle sizes are bigger, it is more difficult to put them into suspension, and the time that they stay forming the cloud is shorter, making more difficult their ignition. However, it is important to notice here that the shape and density of biomass products make this phenomenon different from coals, heavier and normally bigger.

4.2 Ignition Sensibility

Ignition sensibility refers to every characteristic of the dust related to its easiness to ignite. The parameters included in this group are used to design the prevention measurements for the facilities, and are three: minimum ignition temperature (MIT), lower explosibility limit (LEL) and minimum ignition energy (MIE).

4.2.1 Minimum Ignition Temperature

The minimum ignition temperature is the lowest temperature at which the ignition of the dust starts. This parameter is determined in two configurations: with the dust dispersed as a cloud (minimum ignition temperature on a cloud) or deposited as a layer (minimum ignition temperature on a layer). The maximum superficial temperatures of the equipment are stablished according to their values, in °C.

The minimum ignition temperature on a layer is determined following the Standard ISO 80079-20-2 [8]. The MITl is essential since it determines the maximum temperature that the surface of any equipment at the facility can present, not reaching this value at any moment to avoid the flammability of the dust.

The test equipment is schematically shown in Fig. 4.2 and consists on a circular electrically heated surface of 20 mm diameter that can reach 400 °C. The temperature of the plate is controlled with two thermocouples located in the centre of the plate, while the temperature in the sample is registered by a thermocouple located 2–3 mm on the plate. A metallic ring of 10 mm diameter and 5 mm thickness (according to the Standard [8]) is located in the middle of the plate, and the sample is deposited inside the ring with a spatula, uniformly distributed without compacting. Once the layer is formed, ignition—no ignition is identified. Ignition occurs if one the following three situations occur in 30 min: (i) incandescence or visible flame is shown, (ii) 450 °C is reached in the sample, or (iii) a 250 K elevation of the plate temperature is measure.

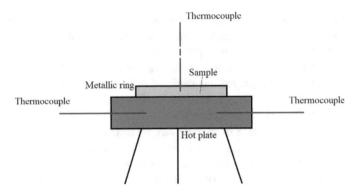

Fig. 4.2 Schematic of minimum ignition temperature on a layer equipment

If ignition is observed, the test should be carried out at a lower temperature. In the contrary, if ignition is not observed, the test is repeated at a higher temperature. Tests have to be carried until a minimum ignition temperature is determine with a difference of 10 K with the first one where ignition does not occur. It has to be checked three times.

Several researchers have worked in the study of this parameter. The smouldering combustion that takes place is controlled by a combination of both the thermal ignition [9] and the diffusion of oxygen [10]. Depending on which one of them is predominant, smouldering will evolve differently, and depending, in order of importance, on the nature of dust, the layer thickness and the particle size [11].

Looking at the nature of dust, the chemical composition of the materials heavily influences their tendency to ignition. On one hand, a higher organic content implies a higher risk of ignition, since more volatiles favours the starting of these processes. However, if the ashes formed during the ignition of the layer form a shell around the dust, it can act in two opposite directions: it can prevent the access of oxygen to the sample, hindering the combustion process, or it can increase the heat inside the layer, favouring it.

Regarding the layer thickness and the particle size, it is well known that, when working with dust materials, both a higher thickness of the dust layer and a decrease on the particle size cause a decrease of the minimum ignition temperature on a layer, favouring its ignition. However, recent studies proved that this behaviour is not the same with bulk materials, and the risk is the same for dust and bulk materials in the case of several biomass [6].

The minimum ignition temperature on a cloud (MITc) is determined following the standard ISO 80079-20-2 [8]. The furnace (Fig. 4.3) consists on a silicon tube, placed vertically and open to the air in its bottom, surrounded by a metallic heated cylinder. Its top is connected to the dust disperser through a gas adapter, and the dust is dispersed in the oven through a solenoid valve that releases compressed air. The procedure consists on dispersing 0.1 g of sample into the equipment pre-heated at 500 °C thanks to an overpressure of 0.1 bar. Ignition is detected visually helped with

Fig. 4.3 Schematic of minimum ignition temperature on a cloud furnace

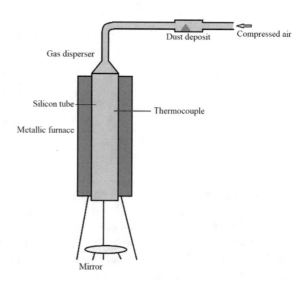

a mirror located at the bottom of the tube, when a visual flame is observed at the bottom of the silicon rube. Sparks without flame are not considered ignition.

If no ignition occurs, the test is repeated increasing the temperature of the furnace 50 K until ignition or until 1000 °C. Once ignition appears, the sample mass and the air pressure are varied until the most vigorous explosion is observed. With the same values of mass and air pressure, temperature has to be reduced 20 K until a temperature where no ignition occurs 10 times. The minimum ignition temperature on a cloud is the lowest temperature of the equipment at which ignition occurred, deducting 20 K for equipment temperatures of more than 300 °C or deducting 10 K for equipment temperatures of less or equal to 300 °C.

This parameter is essential to ensure that any location of the facilities will never reach that temperature when a dust cloud can be formed, avoiding its ignition.

4.2.2 Lower Explosibility Limit

The lower explosibility limit is the lowest concentration of dust dispersed in air that is potentially explosive. We need to ensure that this concentration (or higher) is never dispersed into the air in our facility to avoid explosions.

The lower explosibility limit is determined according to the standard EN 14034-3 [12]. The test equipment is a 20 L sphere, schematically shown in Fig. 4.4.

The 20 L sphere has two standard dispersion systems: a perforated dispersion ring and a rebound nozzle. Following the standards, both of them can be used indistinctively. However, Sanchirico et al. [13] concluded after testing them that the dispersion methods modify the particle size of the dust through its breakage up to a point that the standard procedure may lead to misleading results. Moreover, this two standard

Fig. 4.4 Scheme of 20 L
sphere

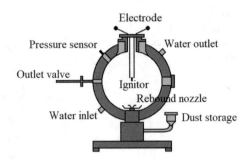

dispersion nozzles cannot be used with particles with higher particle size or irregular shapes, as many biomass materials, since the wholes do not allow this materials to pass. To this end, researchers are developing new options that allow the dispersion of fibrous materials, such as the spherical nozzle dust disperser that University of Leeds uses in their 1 m^3 sphere, additionally to a change on their dust pot from the standard 5 L to a bigger 10 L one [14].

The sample is placed on the dust pot pressurized to 20 bars. It is released inside the spherical chamber, where electrical ignitors of 5 kJ ignite with an ignition delay of (0.6 ± 0.01) s. The test series starts with 500 g/m^3. If ignition occurs, the test is repeated with a decrease of 50% of the tested mass. If no ignition occurs, the test is repeated with an increase of 500 g/m^3 on the tested mass. At the limit where explosion/no explosion is observed, LEL is the first concentration where no explosion is observed.

In general, it is expressed in terms of weight units by air volume unit. It has been observed that for particle sizes of less than 180 μm of diameter, the LEL is independent of the particle size. However, if the diameter is larger, a quick increment on LEL values is observed when increasing the particle size [15], making more difficult the ignition of the cloud.

4.2.3   Minimum Ignition Energy

The minimum ignition energy (MIE) is the lowest energy that ignites the most flammable mixture of a dust cloud, determined in mJ [16].

MIE can be determined using the Hartmann apparatus or using the Mike 3 apparatus. The second one provides results which are equal or lower than the ones measured with the Hartmann apparatus, and it will be the one under consideration in the present manuscript. However, historically the Hartmann tube is the one that has been used for determining this parameter, and the schematics are showed in Fig. 4.5.

The standard regulating the determination of the minimum ignition energy using the Mike 3 apparatus is the ISO 80079-20-2 [8]. The apparatus consists on a vertical cylindrical glass tube with an inner diameter of 68 mm and a height of 300 mm, giving a volume of 1.2 L. Half way up the tube, ignition electrodes are located

Fig. 4.5 Schematic of
Hartmann tube

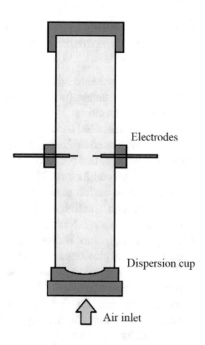

Electrodes

Dispersion cup

Air inlet

with a gap between them of 6 mm. The tube is connected at its base to a 50 mL air reservoir pressurised to 7 bar. The sample is deposited in the bottom of the tube, and it will then be dispersed with an air current thanks to the pressurised air inside the reservoir. The spark is automatically activated with an ignition delay of 120 ms, to allow time for the dispersion of the sample in the tube. Two parameters vary during the tests: the mass of sample studied and the spark energy. A visual verification of a propagating flame in the glass tube is required for ignition to be considered. The MIE is then estimated using the probability of ignition: $\log \text{MIE} = \log E_1 - I[E_2]((\log E_2 - \log E_1))/((NI + I)[E_2] + 1)$, where E_1 is the highest energy at which no ignition occurs, E_2 is the lowest energy at which ignition occurs, I means ignition and NI means no ignition.

4.3 Explosion Severity

The characteristics included in this group allow the evaluation of the explosion consequence. Therefore, they are used to design the protection measurements for the facilities.

This group includes: maximum pressure of explosion (P_{max}), characteristic constant (K_{max}), maximum rate of pressure rise ($(dP/dt)_{max}$) and limiting oxygen concentration (LOC). The first three ones are closely related and determined following the standard EN 14034, part 1 and part 2 [12, 17].

4.3.1 Maximum Pressure, Characteristic Constant and Maximum Rate of Pressure Rise

The maximum pressure is defined as the maximum difference between the pressure at the ignition moment (normal pressure) and the maximum pressure registered at the pressure-time curve.

The equipment used for determining it is the 20 L sphere that was detailed before. The sample is placed in the dust pot, pressurized to 20 bars. It is then released into the vessel and two igniters 5 kJ are ignited 6 ms after the opening of the valve, trying to ignite the dust cloud formed. The pressure inside the vessel is continuously recorded vs time. The explosion pressure is the arithmetic mean of the values measured by pressure sensors. The first test is developed with 250 g/m^3 sample and repeated for a dust concentration interval, with increments of 250 g/m^3 or decrements of 50%. The explosion pressure is determined for all these concentration, and the maximum value of these explosions is considered the maximum pressure.

The rate of explosion pressure rise is the maximum value of pressure rise per unit of time during an explosion. The data used is the same obtained by the experimental procedure described for the maximum pressure of explosion, and the rate of explosion pressure is determined as the slope of the curve that goes from atmospheric pressure to the maximum pressure.

The characteristic constant is used to classify the explosion type as shown in Table 4.1. It is independent of the volume and determined with the equation: $(dP/dt)_{max} \cdot V^{(1/3)} = k_{max}$.

P_{max} of a dust-air mixture is sensitive to flow properties. When the effect of turbulence is not taken into account, the theoretical approach may lead to underestimations or over estimations of the design strength of industrial equipment [18].

$(dP/dt)_{max}$ is the maximum rise of pressure, and it mainly depends on the combustion speed. Its graphical determination is shown in Fig. 4.6.

The explosion violence $(dP/dt)_{max}$ is strongly affected by the particle diameter, while maximum explosion pressure is not so dependent upon the particle size.

The slow decrease of P_{max} with the increase of the particle size means that small changes in the generated volatile content are verified, and it is also an indication that non adiabatic effects can increase with the rise of particle dimensions.

For high dust loadings, conditions when the maximum for the explosions parameters are obtained, devolatilization process cannot be completed within the flame

Table 4.1 Explosion type depending on Kst

Explosion type	k$_{max}$ (m·bar/s)	Explosion
St 0	0	No explosion
St 1	1–200	Weak explosion
St 2	200–300	Strong explosion
St 3	>300	Very strong explosion

Fig. 4.6 Typical pressure
data for explosions

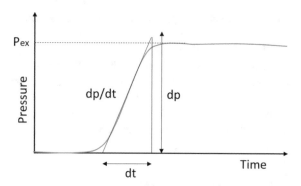

front, and the larger is the particle, the greater is the undevolatilized dust mass fraction. This increase, plus the increase of the residual char makes no contribution to the flame propagation process and begins to absorb a larger fraction of the flame released heat flux. As a result, the effective heating flux driving the fresh pyrolysis wave is diminished.

The strong dependency of the dP/dt is related to the strong decrease verified in the devolatilization rate when the particle size is raised.

This behaviour affects the combustion rate, and a strong decrease in the burning velocity is verified. The subsequent increase of the combustion time leads to the increase of heat losses to the exterior of the reactor, contributing for a reduction of the P_{max} [19].

At elevated concentrations, the burning velocity is high but the depth of penetration of the combustion wave decreases and becomes less than the average particle radios. The excess undevolatilized fraction mass plus the residual char makes no contribution to the flame propagation process; it begins to absorb a large fraction of the flame heating flux, and the effective heating flux, that drives the fresh pyrolysis wave ahead of the flame front, is diminished. As the particle size increases still further, the pyrolysis wave progresses more slowly into each particle and this effect is accentuated with the increase of particle diameter. As a result, there is a smaller contribution of volatiles from each particle that is compensated for by the higher surface area of the particles [15].

4.3.2 Limiting Oxygen Concentration

The limiting oxygen concentration is the minimum percentage of oxygen at which an explosion occurs, in volume percent. The LOC is experimentally determined with the 20 L sphere as follows: the inner volume is filled with the inert gas/air mixture needed to ensure that the oxygen concentration is the one to be tested. The sample is located into the dust pot pressurised at 20 bars. It is then dispersed into the vessel, forming a dust cloud and the igniters are activated. Explosion/no explosion is observed. If an explosion occurs, the oxygen concentration has to be reduced by 1%.

The LOC measurement is not normally used directly to provide inerting levels as a reasonable safety factor should be applied to account for the sensitivity, accuracy and reliability of the plant monitoring system to establish safe inerting levels in industrial processes and to set oxygen concentration alarms or interlocks in internal vessels [20]. NFPA 69 [21] recommends keeping the system oxygen concentration at least 2% lower than measured LOC value when protecting equipment.

4.4   Thermal Susceptibility

This group is used to know the thermal behaviour of solids and to determine their self-combustion tendency. The main parameters included in this group are: Maciejasz index (MI), temperature of emission of flammable volatiles (TEV), and those determined by thermogravimetric and differential scanning calorimetric tests.

4.4.1   Maciejasz Index

The Maciejasz index is determined by the reactivity of organic products to hydrogen peroxide, reflecting their susceptibility to self-ignition. The experimental procedure is developed with the equipment schematically shown in Fig. 4.7 and the procedure that follows. 10 g of sample mixed with 10 cm^3 of distilled water is introduced into a Dewar flask, and 30 mL of hydrogen peroxide is slowly added from a burette and

Fig. 4.7 Schematic of Maciejasz index equipment. Adapted from [22]

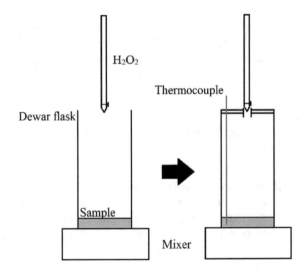

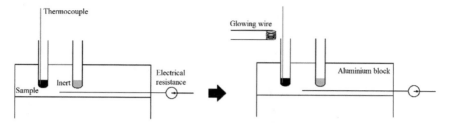

Fig. 4.8 Schematic of temperature of flammable volatiles equipment. Adapted from [22]

continuously agitated, measuring the time. The MI is determined as MI = 100/t, where t is the time in minutes that the sample needs to rise its temperature 65 K. Self-ignition is considered when MI is higher than 10, and the highest this number, the quickest the reaction is.

4.4.2 Temperature of Flammable Volatiles

The temperature at which flammable volatile substances are released during the thermal degradation of organic matter is called the temperature of flammable volatiles. Flammable gases can be emitted when heating up a solid organic material, and these gases are normally more sensitive to ignition than the solid, so they normally ignite at lower temperatures. To determine this temperature, a portion of the sample is heated at increasing temperatures, while approaching a glowing wire. If no flame appears, the temperature of the sample is increased by 10 K and the process is repeated. When a flame appears, the temperature is reduced by 10 K and the process is repeated. The temperature of emission of flammable volatile is established as 10 K below the temperature at which a flame appears (Fig. 4.8).

4.4.3 Thermogravimetric Analysis

In Thermogravimetric Analysis (TG) the weight of the sample is measured when heated following an established heating rate and flowing a process gas. Figure 4.9 shows a typical TG plot, with the derivative curve dTG (sometimes called DTA: Differential Thermal Analysis) superimposed.

Depending on the heating rate and the process gas, different reactions and processes can be simulated such as combustion, pyrolysis, carbonization, crystallization, etc. Typical process gas are air, oxygen or inert gasses; while heating rates may vary from 1 to 100 K/min. Thermogravimetric analysis provides a wide range of information if the curves are treated properly. The most immediate parameters defined in TG curve are moisture content and onset temperature or induction temperature.

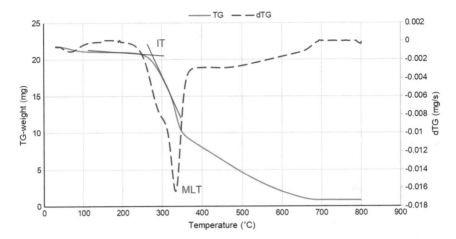

Fig. 4.9 Typical thermogravimentric curve

The moisture content can be calculated using the initial mass, and the weight lost at 105 °C. The onset temperature represents the temperature at which the sample starts rapidly losing weight which means that the material starts disintegrate and can be considered as a thermal stability indicator. On the other hand, dTG curve provides the maximum loss of weight temperature, which usually represents the devolatilization temperature.

TG analyses have been widely used to study the kinetics of combustion processes [23, 24]. The main advantages of this analysis are its rapid assessment of the value and the temperatures at which combustion starts and ends [25]. Thanks to the thermogravimetric curves, the three stages of a combustion process can be easily determined [26]. First of all, the dehydration and drying process, followed by the devolatilization of the organic matter, where the main weigh of loss occurs. Finally, the inorganic material of the sample is decomposed by the oxidation of the char remaining after the volatiles were removed from the samples. Added to this, TG is an excellent method for accurate determination of the ignition temperature of solid fuels [27]. There are also some researchers that have developed a method to obtain "self-ignition potential" by performing the test at different heating rates [28].

As a main disadvantage of these test, we can mention the size of the tested samples. Only small amounts of finely divided dust can be test, being sometimes not representative of the sample. However, it gives an accurate and essential description and a first approach to the study.

Depending on the samples studied and due to their structural composition, two points of maximum weight loss can be obtained when performing the test with air or inert gas flow: the resultant from the devolatilization of light volatiles (MLT_LV) and the corresponding to the devolatilization of heavy volatiles (MLT_HV). As most of the biofuels have a holocellulose composition, it is possible to relate each temperature to the main volatile matter: light volatiles will correspond to holocellulose content, and heavy volatiles will represent lignin content. The remaining mass would correspond to the final residue, which is usually char. Figure 4.10 shows those peaks.

Some authors have carried out the deconvolution of the DTG curve, so it is possible to estimate the composition of the main components by calculating the area beneath the curve that represents that component [29, 30].

If the test is performed substituting the air or inert gas stream by an oxygen one, the reaction is quicker. In this scenario, the sample has a sudden loss of weight at a temperature defined as the characteristic oxidation temperature (Fig. 4.11). This

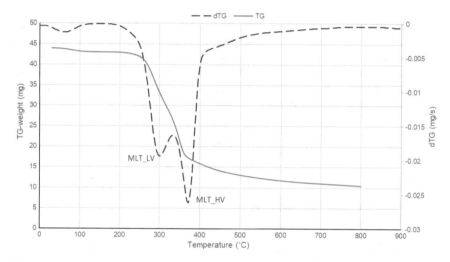

Fig. 4.10 Thermogravimetric curve for biomasses

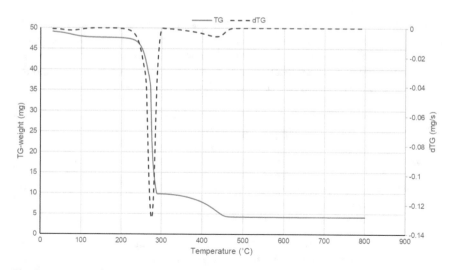

Fig. 4.11 Curve of thermogravimetry with oxygen test

Table 4.2 Self-ignition risk depending on characteristic temperature values

Self-ignition risk	Characteristic temperature (°C)
Very high	≤ 250
High	250–299
Medium	300–349
Low	≥ 350

temperature allows a classification of the samples regarding its self-ignition risk as shown in Table 4.2.

Using data from conventional thermogravimetry analysis with air stream and the mathematic model proposed by Cumming [31], the apparent activation energy of a sample is calculated at the temperature of maximum weight loss of holocellulose.

Cumming's equation relates the apparent activation energy to the rate of weight loss, allowing the estimation of the "apparent activation energy", E_a, from the slope of the least-squares straight line fitted to the chosen test data:

$$ln\left(-\frac{1}{W_\infty} \cdot \frac{dw}{dt}\right) = -\ln A - \frac{E_a}{R \cdot T}$$

where:

W_∞ Weight of unburnt sample (g)
dw/dt Instant velocity of weight loss (g/s)
A Frequency factor
E_a Apparent activation energy (J/mol)
R Universal gas factor (8.314 J/mol K)
T Absolute temperature (K).

However, thermodynamic and mathematic basis of Cumming's equation, is based on the equations that characterise the first-order reactions, which simplifies and makes easier the calculus and use of the value of the apparent activation energy, but is not accurate enough when other reactions take place. Using Arrhenius equation and the thermochemical conversion rate equation, is it possible to define kinetics as:

$$\frac{d\alpha}{dT} = \frac{A}{\beta} \cdot f(\alpha) \cdot e^{-\frac{E}{RT}}$$

where:

α normalized conversion degree
β heating rate (K/min)
$f(\alpha)$ reaction model conversion function.

The estimation of the kinetic parameters is carried out using model-fitting methods, or model-free methods. In the first ones, $f(\alpha)$ is assumed fitting the data to the model, and the remaining parameters are calculated. On the other hand, model-free methods are more complex as they assume that activation energy does not remain constant during reaction and use iso-conversional techniques, but the estimation is more accurate [32]. Some kinetic models and their corresponding $f(\alpha)$ can be shown in Table 4.3.

The value of activation energy (E_a) has been used to classify coal samples according to their self-ignition risk as shown in Table 4.4.

When plotting the values of Ea and Tcaract a thermal susceptibility graph is obtained in which different degrees of reactivity or tendency to spontaneous combustion can be distinguished (Fig. 3.3).

Table 4.3 Parameters used for different kinetic models [33]

Kinetic model	$f(\alpha)$	$f'(\alpha)$	$g(\alpha)$
Phase boundary controlled reaction (contracting area)	$2(1-\alpha)^{1/2}$	$-\dfrac{1}{(1-\alpha)^{1/2}}$	$1-(1-\alpha)^{1/2}$
Phase boundary controlled reaction (contracting volume)	$3(1-\alpha)^{2/3}$	$-\dfrac{2}{(1-\alpha)^{1/3}}$	$1-(1-\alpha)^{1/3}$
Random nucleation followed by an instantaneous growth of nuclei (Avrami-Erofeev, n $= 1$)	$1-\alpha$	-1	$-\ln(1-\alpha)$
Random nucleation followed by an instantaneous growth of nuclei (Avrami-Erofeev, n $\neq 1$)	$(1-\alpha)[-\ln(1-\alpha)]^{1-1/n}$	$n+n\cdot\ln(1-\alpha)-\dfrac{1}{[-\ln(1-\alpha)]^{1/n}}$	$[-\ln(1-\alpha)]^{1/n}$
2D diffusion	$\dfrac{1}{[-\ln(1-\alpha)]}$	$-\dfrac{1}{(1-\alpha)[-\ln(1-\alpha)]^2}$	$\alpha+(1-\alpha)\ln(1-\alpha)$
3D diffusion (Jander)	$\dfrac{3(1-\alpha)^{2/3}}{2[1-(1-\alpha)^{1/3}]}$	$\dfrac{\frac{1}{2}-(1-\alpha)^{-1/3}}{[1-(1-\alpha)^{1/3}]^2}$	$\left[1-(1-\alpha)^{1/3}\right]^2$
3D diffusion (Ginstling-Brounshtein)	$\dfrac{3}{2}\left[(1-\alpha)^{1/3}-1\right]$	$\dfrac{(1-\alpha)^{4/3}}{2\left[(1-\alpha)^{-1/3}-1\right]^2}$	$1-2/3\alpha-(1-\alpha)^{2/3}$

Table 4.4 Self-ignition risk depending on activation energy values

Self-ignition risk	Characteristic temperature (°C)
Very high	≤ 79
High	80–89
Medium	90–94
Low	≥ 95

4.4.4   Differential Scanning Calorimetry

In the case of Differential Scanning Calorimetry (DSC) the sample is heated at a regular rate, previously established, and a reference inert product is placed in another crucible. The difference in temperature between the sample and the reference is measured and recorded against the temperature of the oven and the exchanges of heat in the sample are determined. Figure 4.12 shows a typical DSC record. The parameters used to characterize different substances are the minimum temperature at which the exothermic reaction begins (initial temperature, IET), the maximum temperature reached during the exothermic reaction (final temperature, FET) and the temperature at which the fast exothermic reaction commences (change of slope temperature, CST). If the area of the curve is calculated, it is also possible to obtain the flow energy for the exothermic process, which corresponds to the positive peak of the curve, and the endothermic process, which corresponds to the negative peak of the curve. It can be used as a supplementary technique with TG, so not only the combustion temperature is obtained, but also the combustion heat [34].

DSC can be used to obtain the calorific requirement, which is the total heat needed to increase temperature, so the pyrolysis peak is achieve [35], and to define the total energy required to carry out a thermal decomposition [36] that means that DSC can be used to define thermal parameters used in industry processes. The heat peaks also might represent phase transitions and physical transformations, and the enthalpies of these transitions, so a better knowledge of the thermal behaviour of a sample can be obtained. When testing lignocellulosic biomass, it provides further information to estimate the amount of cellulose, hemicellulose and lignin, as the thermal degradation of each component is also represented in DSC curves [37].

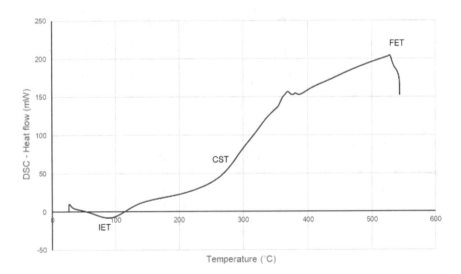

Fig. 4.12 Typical curve of differential scanning calorimetry

Furthermore, it is possible to obtain specific heat capacity using the following equation [38]

$$\frac{dH}{dt} = m \cdot c_p \cdot \frac{dT}{dt}$$

where:

m is the mass of the sample

c_p is the specific heat capacity.

References

1. Traoré M, Dufaud O, Perrin L, Chazelet S, Thomas D (2009) Dust explosions: how should the influence of humidity be taken into account? Process Saf Environ Prot 87(1):14–20
2. Yuan J, Wei W, Huang W, Du B, Liu L, Zhu J (2014) Experimental investigations on the roles of moisture in coal dust explosion. J Taiwan Inst Chem Eng 45(5):2325–2333
3. Cashdollar KL (1996) Coal dust explosibility. J Loss Prev Process Ind 9(1):65–76
4. Calle S, Klaba L, Thomas D, Perrin L, Dufaud O (2005) Influence of the size distribution and concentration on wood dust explosion: experiments and reaction modelling. Powder Technol 157(1):144–148
5. Eckhoff RK (2009) Understanding dust explosions. The role of powder science and technology. J Loss Prev Process Ind 22(1):105–116
6. Fernandez-Anez N, Garcia-Torrent J (2019) Influence of particle size and density on the hot surface ignition of solid fuel layers. Fire Technol 55(1):175–191
7. Restuccia F, Fernandez-Anez N, Rein G (2019) Experimental measurement of particle size effects on the self-heating ignition of biomass piles: homogeneous samples of dust and pellets. Fuel 256:115838
8. ISO/IEC DIS 80079-20-2. Explosive atmospheres-Part 20-2: Material characteristics-Combustible dusts test methods
9. Bowes PC, Townshend SE (1962) Ignition of combustible dusts on hot surfaces. Br J Appl Phys 13(3)
10. Nelson M (1995) Detection and extinction of fire and smouldering in bulk powder
11. Miron Y, Lazzara CP (1988) Hot-surface ignition temperatures of dust layers. Fire Mater 12:115–126
12. EN 14034-2:2006+A1:2011. Determination of explosion characteristics of dust clouds-Part 2: determination of the maximum rate of explosion pressure rise (dp/dt)$_{max}$ of dust clouds
13. Sanchirico R, Di Sarli V, Russo P, Di Benedetto A (2015) Effect of the nozzle type on the integrity of dust particles in standard explosion tests. Powder Technol
14. Medina CH et al (2015) Comparison of the explosion characteristics and flame speeds of pulverised coals and biomass in the ISO standard 1 m^3 dust explosion equipment. Fuel 151:91–101
15. Pilão R, Ramalho E, Pinho C (2002) Influence of particle size on the explosibility of air/cork dust mixtures. In: Proceedings of 9th Brazilian congress of thermal engineering and sciences (CIT02-0689. pdf), pp 15–18
16. Eckhoff RK (2002) Minimum ignition energy (MIE)—a basic ignition sensitivity parameter in design of intrinsically safe electrical apparatus for explosive dust clouds. J Loss Prev Process Ind 15(4):305–310
17. EN 14034-1:2004+A1:2011. Determination of explosion characteristics of dust clouds-Part 1: determination of the maximum explosion pressure p$_{max}$ of dust clouds

18. Dahoe AE, van der Nat K, Braithwaite M, Scarlett B (2001) On the sensitivity of the maximum explosion pressure of a dust deflagration to turbulence. KONA Powder Part J 19:178–196
19. Pilão R, Ramalho E, Pinho C (2006) Overall characterization of cork dust explosion. J Hazard Mater 133(1):183–195
20. Mittal M (2013) Limiting oxygen concentration for coal dusts for explosion hazard analysis and safety. J Loss Prev Process Ind 26(6):1106–1112
21. NFP & Association (2008) Standard on explosion prevention systems (NFPA 69). Quincy, MA, National Fire Protection Association
22. Ramírez Á, García-Torrent J, Tascón A (2010) Experimental determination of self-heating and self-ignition risks associated with the dusts of agricultural materials commonly stored in silos. J Hazard Mater 175(1):920–927
23. Hadden R, Rein G (2011) Burning and suppression of smouldering coal fires. In Coal and peat fires: a global perspective, pp 317–326
24. Eckhoff RK (2005) Explosion hazards in the process industries
25. Magdziarz A, Wilk M (2013) Thermogravimetric study of biomass, sewage sludge and coal combustion. Energy Convers Manag 75:425–430
26. Magdziarz A, Werle S (2014) Analysis of the combustion and pyrolysis of dried sewage sludge by TGA and MS. Waste Manag 34(1):174–179
27. Chen Y, Mori S, Pan W-P (1996) Studying the mechanisms of ignition of coal particles by TG-DTA. Thermochim Acta 275(1):149–158
28. Janković B, Manić N, Stojiljković D, Jovanović V (2020) The assessment of spontaneous ignition potential of coals using TGA–DTG technique. Combust Flame
29. Janković B, Manić N, Dodevski V, Popović J, Rusmirović JD, Tošić M (2019) Characterization analysis of Poplar fluff pyrolysis products. Multi-component kinetic study. Fuel 238:111–128
30. Manić NG, Janković B, Stojiljković DD, Jovanović VV, Radojević MB (2018) TGA-DSC-MS analysis of pyrolysis process of various agricultural residues. Therm Sci 2018:1–15
31. Cumming JW (1984) Reactivity assessment of coals via a weighted mean activation energy. Fuel 63(10):1436–1440
32. Radojevic M, Balac M, Jovanovic V, Stojiljkovic D, Manic N (2018) Thermogravimetric kinetic study of solid recovered fuels pyrolysis. Hem Ind 72(2):99–106
33. Sronsri C, Boonchom B (2018) Thermal kinetic analysis of a complex process from a solid-state reaction by deconvolution procedure from a new calculation method and related thermodynamic functions of $Mn_{0.90}Co_{0.05}Mg_{0.05}HPO_4 \cdot 3H_2O$. Trans Nonferrous Metals Soc China 28(9):1887–1902
34. López-González D et al (2015) Combustion kinetic study of woody and herbaceous crops by thermal analysis coupled to mass spectrometry. Energy 90:1626–1635
35. Nyakuma BB, Johari A, Ahmad A, Abdullah TAT (2014) Comparative analysis of the calorific fuel properties of Empty Fruit Bunch fiber and briquette. Energy Procedia 52:466–473
36. Torabi M et al (2016) We are IntechOpen, the world's leading publisher of Open Access books Built by scientists, for scientists TOP 1%. Intech, vol i, no tourism, p 13
37. Bryś A et al (2016) Wood biomass characterization by DSC or FT-IR spectroscopy. J Therm Anal Calorim 126(1):27–35
38. Collazo J, Pazó JA, Granada E, Saavedra Á, Eguía P (2012) Determination of the specific heat of biomass materials and the combustion energy of coke by DSC analysis. Energy 45(1):746–752

Chapter 5
Composition and Characteristics

The composition of solid substances has a great impact into their flammability and ignition risks. Several efforts have been focused on relating these factors to facilitate the characterisation of materials depending on their composition. To this end, both elemental and numerical analyses are discussed in this chapter, and their correlations with some of the basic flammability parameters of coal and biomass are explained. The chemical composition of the materials has a high effect in their flammability and ignition risks. In the case of coals, it is well known that an increase on the volatiles content, meaning a decrease in the ash content, causes a rise in the flammability probability of the samples.

5.1 Elemental Analysis

Elemental analysis provides the content of carbon, hydrogen, nitrogen and sulphur of the studied samples.

According to the European Union legislation, carbon, hydrogen and nitrogen contents are determined following the standard EN 15407 [1], based on the complete oxidation of the materials. The procedure starts with the conversion of C, H and N in their corresponding gases during the combustion in an atmosphere of oxygen of the sample (CO_2, H_2O and NO_x respectively). The other combustion products are cleaned before the detection of the gases to avoid interferences. NO_x gases are reduced to N_2, and this is determined like the CO_2 and H_2O by infrared spectrometry and thermal conductivity

The total sulphur content is determined in solid fuels mainly for environmental and technical reasons. It is determined following the Standard EN 15408 [2] consisting of the combustion at high temperature on a tube furnace and the detection by infrared absorption. The procedure starts by the burning in the tube furnace of a portion of the sample, with the presence of oxygen at least at 1350 °C. During the combustion, sulphur and its compounds are decomposed and oxidized to SO_2. This SO_2 is detected by the infrared spectrometer.

© The Author(s), under exclusive license to Springer Nature Switzerland AG 2020
N. Fernandez-Anez et al., *Explosion Risk of Solid Biofuels*,
SpringerBriefs in Energy, https://doi.org/10.1007/978-3-030-43933-0_5

Dust chemistry influences both thermodynamics (amount of heat liberated during combustion) and kinetics (rate at which the heat is liberated).

The chemical composition of biomass can, by itself, provide information about the expected behaviour of fuels. For coals, it is related with the rank of the samples. A decrease of the rank means an increase on the oxygen and hydrogen content, due to the presence of shorter chain hydrocarbons, and it implies an increase on the flammability risks of the dusts. There is also a positive correlation observed between the carbon and hydrogen contents with the self-ignition tendency of coal [3].

Additionally, by studying the process of combustion between the samples, more information can be obtained. If we simplify the formula of any studied fuel as CH_yO_z, the combustion reaction in air is the following:

$$CH_yO_z + \alpha(O_2 + 3.76\,N_2) \rightarrow CO_2 + y/2\,H_2 + 3.76\alpha\,N_2$$

where α is obtained by $\alpha = 1+y/4-z/2$ and the number of air moles is $\alpha/0.21$.

Coefficients y and z represent the atomic ratios H/C and O/C, respectively. The balanced combustion equation in air leads to determine the stoichiometric air to fuel mass ratio, A/F:

$$[A/F]_{stoic} = (\alpha/0.21 \times 28.84)/(12 + y + 16z)$$

where 28.84 g per mole is the typical molecular mass of air. Also, dust concentration for stoichiometric concentrations (equivalence ratio $\Phi = 1$) can be obtained for an air density of $1200\ g/m^3$.

The air to fuel ratio is the mass ratio of air to a solid present in a combustion process (Table 5.1).

A graphical way of showing the ratios of chemical components is the Van Krevelen diagram, the 2-D representation of H/C ratios vs O/C ratios. Van Krevelen [7] developed the diagrams to classify coals and predict its compositional evolution during thermal maduration. Tissot et al. [8] modified this diagram in the 1970s to identify the kerogen of sedimentary rocks based on the maturity tracks from pyrolysis data, substituting the atomic H/C ratio by the hydrogen index and the atomic O/C ratio by the oxygen index. This diagram can be used to define an organic sample, and therefore it has a close relation with their thermal behaviour (Fig. 5.1).

Comparison of biofuels with fossil fuels shows clearly that the higher proportion of oxygen and hydrogen, compared with carbon, reduces the energy value of a fuel, due to the lower energy contained in carbon-oxygen and carbon-hydrogen bonds, than in carbon-carbon bonds [9].

Focusing in the first part of the diagram, the coal samples without biomass, the samples located nearby the origin are the most evolved ones, with higher heating values and less flammability risk. When moving away from the origin, the carbon content of the sample decreases, presenting more linear molecules at the same time that the flammability parameters present more risk for the samples. At the end of the diagram, the biomass group represents a large part, inside where the biomasses are categorised in a different way: samples up and left in the area of biomass are

Table 5.1 Chemical parameters for different biomass and coal samples from literature

Sample	H/C	O/C	A/F	$\Phi = 1$	Reference
Wood	1.43	0.59	6.4	188	[4]
Herbaceous	1.49	0.67	5.9	204	[4]
Straw	1.48	0.66	5.9	202	[4]
Shells	1.51	0.63	6.2	194	[4]
Animal biomass	1.51	0.29	9.3	129	[4]
Sludge	1.72	0.49	7.5	159	[4]
Lignite	1.03	0.28	8.8	137	[4]
Subbituminous coal	0.90	0.18	9.9	122	[4]
Bituminous coal	0.72	0.09	11.0	109	[4]
Raw Norway spruce	5.6/48.1	36.3/48.1	6.5	184	[5]
Torrefied Norway spruce	5.2/51.6	35.4/51.6	6.7	178	[5]
Kellingley coal	4.1/65.0	5.5/65.0	11.3	106	[5]
Spruce	3.58	1.55	3.83	313	[6]
Cellulose	1.67	0.833	5.12	234	[6]
rapeseed straw	1.88	0.986			[6]
Spanish lignite	1.42	0.826			[6]
Miscanthus	1.62	0.771			[6]
Carbon monoxide	0	0.75	3.45	–	[6]
Wood	1.59	0.731			[6]
Spanish pine	1.63	0.729			[6]
Barley straw	1.68	0.705			[6]
Forest residue	1.53	0.672			[6]
Bark	1.42	0.637			[6]
Sorghum straw	1.45	0.647			[6]
German lignite	1.09	0.450			[6]
Pitch pine	1.46	0.416	8.09	148	[6]

those with a greater tendency to spontaneous combustion. This is because coals and biomass have different mechanisms of self-heating and spontaneous combustion. For biomass, the beginning of the self-heating process is guaranteed due to their high thermal susceptibility, provided that there is sufficient oxygen to start slow oxidation. The progress of this process and its acceleration toward the self-ignition in biomass are more limited to the ability to transmit the heat generated by the reaction, so that higher thermal conductivity promotes faster progress of the reaction. Biomasses having higher H/C values also provide higher calorific values and they are more susceptible to accelerate the process and provide lower oxidation temperatures [4]. Additionally, samples with low O/C ratio, composition more similar to lignin than

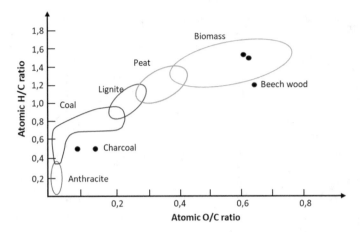

Fig. 5.1 Van Krevelen diagram. Adapted from [7]

to cellulose, present very high tendency to self-ignite the conductivity of lignin is generally higher than that of cellulose [10].

5.2   Proximate Analysis

A proximate analysis of a fuel provides the percentage of the material that burns in a gaseous state (volatile mater), in the solid state (fixed carbon), and the percentage of inorganic waste material (ash). The result of these analyses are essential to determine the possible energetic use of materials. There are several standards, both in Europe and America, which need to be follow to determine the composition of fuels, depending on the type of fuels that are tested. There is a list of the ones considered of interest in the present manuscript in Table 5.2.

Here, we are going to briefly explain the experimental procedure to determine moisture, ash and volatile content following the EU's standards ISO 18134 [11], ISO 18122 [12] and ISO 18123 [13], respectively. Fixed carbon is determined as the complementary part to sum 100%.

5.2.1   Ash Content

First of all, a laboratory dish is weighted. Between 1 and 2 g of sample is placed on it and it is weighted empty (Fig. 5.2). The laboratory dish with the sample is placed inside the oven at room temperature.

The temperature of the oven has to rise under a determinate range, depending on the sample that is tested:

Table 5.2 Standards used for proximate analysis

	Material	Standard	Standard that needs to be referenced
Ash	Solid biofuels	UNE-EN ISO 18122	UNE EN ISO 18122 : 2016. Solid biofuels - Determination of ash content
	Hard coal and anthracite	UNE 32111	UNE 32111: 1995. Hard coal and anthracite. Determination of ash content with automatic programmable furnaces
	Solid mineral fuels	UNE 32004	UNE 32004:1984. SOLID Mineral fuels - Determination of ash
	Solid recovered fuels	UNE-EN 15403	UNE-EN 15403:2011. Solid recovered fuels - Determination of ash content
	Plastics	UNE- EN ISO 3451-1	UNE-EN ISO 3451-1:2008. Plastics - determination of ash - part 1: general methods
		ASTM D5630-13	ASTM D5630-13. Standard Test Method For Ash Content In Plastics
	Graphite sample	ASTM C561-16	ASTM C561-16. Standard Test Method for Ash in a Graphite Sample
	Coal tar and pitch	ASTM D2415-15	ASTM D2415-15. Standard Test Method for Ash in Coal Tar and Pitch
	Petroleum products	ASTM D482-19	ASTM D482-19 Standard Test Method for Ash from Petroleum Products
	Petroleum coke	ASTM D4422-19	ASTM D4422-19. Standard Test Method for Ash in Analysis of Petroleum Coke
	Activated carbon	ASTM D2866-11	ASTM D2866-11. Standard Test Method for Total Ash Content of Activated Carbon

(continued)

Table 5.2 (continued)

	Material	Standard	Standard that needs to be referenced
Volatiles	Solid biofuels	UNE-EN 18123	UNE-EN 18123:2016. Solid biofuels - Determination of the content of volatile matter (ISO 18123:2015)
	Hard coal and coke	UNE 32019	UNE 32019:1984. Hard coal and coke. Determination of volatile matter content.
	Solid recovered fuels	UNE-EN 15402	UNE-EN 15402:2011. Solid recovered fuels - Determination of the content of volatile matter
	Plastics	UNE 53272	UNE 53272:2019. Plastics. Determination of volatile content
	Particulate wood fuels	ASTM E872-82	ASTM E872-82 (2019). Standard Test Method for Volatile Matter in the Analysis of Particulate Wood Fuels
	Activated carbon samples	ASTM D5832-98	ASTM D5832-98(2014) Standard Test Method for Volatile Matter Content of Activated Carbon Samples
	Coke and coal	ASTM D3175-20	ASTM D3175-20 Standard Test Method for Volatile Matter in the Analysis Sample of Coal and Coke
Moisture	Solid biofuels	UNE-EN ISO 18134	UNE-EN ISO 18134:2016. Solid biofuels - Determination of moisture content - Oven dry method - Part 3: Moisture in general analysis sample (ISO 18134-3:2015)
	Hard coal and anthracite	UNE 32001	UNE 32001:1981. Hard coal and anthracite. Determination of total moisture

(continued)

Table 5.2 (continued)

Material	Standard	Standard that needs to be referenced
Solid mineral fuels	UNE 32002	UNE 32002:1995. Solid mineral fuels. Determination of moisture in the analysis sample.
Plastics	UNE-EN ISO 585	UNE-EN ISO 585:2000. Plastics - unplasticized cellulose acetate - determination of moisture content (iso 585:1990)
Particulate wood fuels	ASTM E871-82	ASTM E871-82 (2019) Standard Test Method for Moisture Analysis of Particulate Wood Fuels
Coal	ASTM D3302/D3302M-17	ASTM D3302/D3302M-17. Standard Test Method for Total Moisture in Coal
Graphite sample	ASTM C562-15	.

Fig. 5.2 Crucible for ash content determination

Temperature rises from room temperature to 250 °C in 30–50 min and it remains at this temperature for 60 min. Later, it rises to 550 ± 10 °C in 30 min and it remains for 120 min.

When the heating time finishes, the laboratory dish with the ashes is removed from the oven and cooled for 10 min outside and for 15 min inside a desiccator with a dehydrating agent.

Ash content is determined with the following equation:

$$A_d = (m_3 - m_1)/(m_2 - m_1) \times 100 \times 100/(100 - M_{ad})$$

where

m_1 is the laboratory dish weight, in grams
m_2 is the laboratory dish and the sample weight before heating, in grams
m_3 is the laboratory dish and the ash weight after heating, in grams
M_{ad} is the moisture of the sample.

5.2.2 Volatile Content

The oven is pre-heated at $900 \pm 10\ °C$. Before the test procedure, furniture must be cleaned by placing the crucibles in a rack (Fig. 5.3), and inside the oven for 7 min. They are removed from the oven and cooled first in a flat surface and later in the desiccator. When it is cold, the crucible and its cover are weighted. Between 1 and 1.01 g of sample are placed inside the crucible and the crucible, its cover and the sample are weighted.

Fig. 5.3 Crucible for volatile content

The crucible is placed in the rack uncovered and the crucible, the rack and the cover are introduced into the oven for 7 min. After this time, they are removed and cooled. When they are cold, the crucible, its content and its cover are weighted.

Volatile content is determined with the following equation:

$$V_d = ((100(m_2 - m_3))/(m_2 - m_1) - M_{3d}) \times (100/(100 - M_{3d}))$$

where

m_1 is the crucible weight, in grams
m_2 is the crucible and the sample weight before heating, in grams
m_3 is the crucible and its content after heating, in grams
M_{ad} is the moisture of the sample.

5.2.3 Moisture Content

Before the test, the weighing bottle is dried in an oven at 105 ± 2 °C.

The oven is preheated at 105 ± 2 °C. The weighing bottle and its cover (Fig. 5.4) are weighted. Between one and two grams of sample are placed inside the weighing bottle and this with the cover are weighted.

The weighing bottle is placed in the oven at 105 ± 2 °C discovered for at least 60 min. Because of the heterogeneity of the biomass samples, they have been placed for at least 2 h, after that no mass change is observed after future drying. After this time, they are removed and cooled. When they are cold, the weighing bottle, its content and its cover are weighted.

Fig. 5.4 Weighing bottle for moisture content

Moisture content is determined with the following equation:

$$M_{ad} = (m_2 - m_3)/(m_2 - m_1) \times 100$$

where

m_1 is the weighing bottle and its cover weight, in grams
m_2 is the weighing bottle, its cover and the sample weight before drying, in grams
m_3 is the weighing bottle, its cover and the sample weight after drying, in grams.

5.2.4   Influence of Elemental Analyses in the Flammability of Fuels

The chemical composition of fuels has been pointed out as one of the main parameters influencing their thermal behaviour for a long time.

For coals, it was determined that a higher degree of maturity implies a lower risk of ignition. This is why it has always been accepted that the coals with higher carbon contents and lower H/C ratios (anthracite and bituminous coal) are the ones with lower flammability risk.

Anthracite and most of the semi anthracite coal particles ignited heterogeneously while bituminous coal particles ignited homogenously in the gas phase. As the coal rank decreases to lignite, the ignition mode seems to switch to a mixed heterogeneous/homogeneous behaviour. The increase of fragmentation tendency in the case of lignite influences such behaviour. If fragmentation happens before ignition, the fragments of the lignite ignite heterogeneously. However, if fragmentation happens during or after ignition, then mixed homogenous/heterogeneous ignition is likely to occur. Biomass particles appeared to ignite homogeneously forming faint spherical flames [14, 15].

In previous work [16], we have observed a relation between the self-ignition tendency of different biomass, through the oxidation temperature, and the oxygen content, as well as a negative one with the A/F ratio.

However, researches show that biomass materials do not follow the same trend. Garcia Torrent et al. studied the influence of the composition of biomass in their flammability behaviour, and observed that samples with higher H/C ratios and lower O/C ratios were the products leading to lower oxidation temperatures, so they are more reactive.

The explanation of this phenomenon could be linked to different mechanisms of self-heating development and spontaneous combustion, since for biomasses the beginning of the self-heating process is guaranteed by their high thermal susceptibility, provided that there is sufficient oxygen to start slow oxidation pro-cesses. The progress of this process and its acceleration toward the self-ignition in biomass are more limited to the ability to transmit the heat generated by the reaction, so that higher thermal conductivity promotes faster progress of the reaction. Biomasses having higher H/C values also provide higher calorific values and they are more susceptible to accelerate the process and provide lower oxidation temperatures. In any case, it should not be forgotten that the initiation and development of the ignition process is conditioned by heat generation due to oxidation and the ability to dissipate the heat. Here, not only conductivity is important but also the oxygen ability to access to the material surface [4].

These two different behaviours for coals and biomass have been observed also for volatiles content. When studying the dependence of minimum ignition temperature of different materials with the volatiles content, Rybak et al. [17] observed that coal's minimum ignition temperature linearly decreases with the volatiles content of the samples. However, when studying biomass, this tendency changes to an increasing, non-linear one.

The equivalence ratio has proved to be an indicator of the behaviour of dust during explosions. It has been observed that the severity of explosions increases with the increase of this ratio, until a point where it starts slowly decreasing.

5.2.5 Influence of Proximate Analyses in the Flammability of Fuels

The influence of volatile and ash contents in the flammability of coals have been widely studied.

An increased on the volatile content (meaning a decrease on the ash content) produces an increase on the easiness of ignition of dust clouds and on the severity of the phenomenon that will take place. Furthermore, the highest the volatiles content of the sample, the highest the inert content needed to avoid an explosion. Palmer proposed the following expression to determine this relationship [19]:

$$S = 1000 - 12500V,$$

where S is the inert dust percentage and V the volatiles content.

However, the effects of volatiles into the ignition of layers. An increase on the temperature needed for igniting a dust layer has been observed when the volatiles content increases, making more difficult the ignition of the layer.

In this case, an increase on the ash content normally means a decrease on the ignition easiness, and for that one of the most used methods to avoid the ignition of dusts is to mix them with inert solids. However, ashes can also create a continuous layer that acts as a shell, both maintaining the heat inside (promoting the ignition) and hindering the access of oxygen (inhibiting the ignition).

Regarding moisture content, Depending on the chemical composition of the dust, the moisture can inhibit or promote the ignition of dusts and the severity of the explosion. On the one hand, the inhibition can be caused by agglomeration, heat sink effect, water vapour inerting or inhibition of mass transfer at particle surface. On the other hand, an increase of the ignition and explosion characteristics could be observed when the water presence leads to violent chemical reaction with the dust or when its adsorption chemically modified the particle surface and improves the species diffusion, which can lead to explosion risks higher than for the dry dust.

However, none of these parameters can, by themselves, explain or predict the behaviour of dusts under these circumstances. Saeed et al. [18] proposed a combination of inert (ash and moisture contents) that proving the influence and showing an increase on the concentration of dust needed for an explosion to occur (MEC) when this sum increases. Additionally, the authors pointed out that, even when this relationship exists for all biomass, it differs on the slope of the curves depending on the group of biomass studied.

References

1. EN 15407 (2011) Solid recovered fuels. Methods of determination of carbon (C), hydrogen (H) and nitrogen (N) content
2. EN 15408 (2011) Solid recovered fuels. Methods for the determination of sulphur (S), chlorine (Cl), fluorine (F) and bromine (Br) content
3. Kaymakçi E, Didari V (2002) Relations between coal properties and spontaneous combustion parameters. Turkish J Eng Environ Sci 26(1):59–64
4. García Torrent J, Ramírez-Gómez Á, Fernandez-Anez N, Medic Pejic L, Tascón A (2016) Influence of the composition of solid biomass in the flammability and susceptibility to spontaneous combustion. Fuel 184
5. Medina CH, Sattar H, Phylaktou HN, Andrews GE, Gibbs BM (2015) Explosion reactivity characterisation of pulverised torrefied spruce wood. J Loss Prev Process Ind 36:287–295
6. Slatter DJF, Huescar Medina C, Sattar H, Andrews GE, Phylaktou HN, Gibbs BM (2013) The influence of particle size and volatile content on the reactivity of HC and HCO chemical and biomass dusts. In: Proceedings of the 7th international seminar on fire and explosion hazards
7. Van Krevelen DW (1950) Graphical-statistical method for the study of structure and reaction processes of coal. Fuel 29(12):269–284
8. Tissot B, Durand B, Espitalie J, Combaz A (1974) Influence of nature and diagenesis of organic matter in formation of petroleum. Am Assoc Pet Geol Bull 58(3):499–506
9. McKendry P (2002) Energy production from biomass (part 2): conversion technologies. Bioresour Technol 83(1):47–54
10. Ramírez Gómez A, Medic Pejic L, Querol Aragón E, Grima Olmedo C, García Torrent J (2012) Assessment of self-combustion risks for solid fuels. In: National symposium on handling and hazards of materials in industry (HANHAZ2012). Sevilla
11. International Organization for Standardization (2016) ISO 18134-1:2016. Solid biofuels—determination of moisture content—oven dry method—part 1: total moisture—reference method
12. International Organization for Standardization (2016) ISO 18122: 2016. Solid biofuels—determination of ash content
13. International Organization for Standardization (2016) ISO 18123: 2016. Solid biofuels—determination of the content of volatile matter
14. Kim RG, Li D, Jeon CH (2014) Experimental investigation of ignition behavior for coal rank using a flat flame burner at a high heating rate. Exp Therm Fluid Sci 54:212–218

15. Khatami R, Levendis YA (2016) An overview of coal rank influence on ignition and combustion phenomena at the particle level. Combust Flame 164:22–34
16. Garcia Torrent J, Fernandez Anez N, Medic Pejic L, Montenegro Mateos L (2015) Assessment of self-ignition risks of solid biofuels by thermal analysis. Fuel 143
17. Rybak W, Moroń W, Ferens W (2019) Dust ignition characteristics of different coal ranks, biomass and solid waste. Fuel 237(October 2018):606–618
18. Saeed MA, Farooq M, Andrews GE, Phylaktou HN, Gibbs BM (2019) Ignition sensitivity of different compositional wood pellets and particle size dependence. J Environ Manage 232(December 2018):789–795
19. Palmer KN (1973) Dust explosions and fires [by] KN Palmer

Chapter 6
Fuel Mixtures

Biofuels have proved the environmental benefits that their use make, and how by using these clean energies and reducing the use of fossil fuels we could reduce pollution and emissions. However, they have a main disadvantage: its lower energy density compared to fossil fuels. These two facts together have been the key of an intermediate solution, the use of mixtures of biofuels and fossil fuels, promoting the appearance of co-firing facilities. The use of these unknown materials carry several risks as its tendency to ignite, which is outlined in the present chapter.

6.1 Types and Uses in Industry

The obvious objective of mixing fuels from different origins is to improve some characteristics without losing other advantages. In the case of mixing coal and a biofuel, cleaner energy with a not so low energy density. Mixtures of solid materials have been studied for several applications. As an example, we can see the mixtures of waste for cement production [1].

For energy purposes, mixtures of solid fuels have been widely used in co-firing or co-combustion operations. Co-firing is defined as the combustion of two types of fuel at the same time. One of the main advantages of co-firing is that the equipment previously used for coal combustion can be used for these mixtures with a much lower economical investment, however, further studies are much needed to be conducted such as improvement in boilers design, materials and combustion technology [2].

Co-firing has been mainly implemented by mixing biomass and coal. Reduction of atmospheric emissions of target pollutants compared to traditionally coal-fired power plants [3], and reduction of particulate matter emissions [4] are clear examples of the benefits of partially substituting coal by biomass. These benefits could be increased even more by using waste materials instead of biomass, resulting in a substantial reduction of the disposed volume and in the safe destruction of toxic organic residues, solving part of the problem of waste disposal [5–7].

© The Author(s), under exclusive license to Springer Nature Switzerland AG 2020
N. Fernandez-Anez et al., *Explosion Risk of Solid Biofuels*,
SpringerBriefs in Energy, https://doi.org/10.1007/978-3-030-43933-0_6

One of the main disadvantages of the addition of these substances to the combustion process is the large amount of moisture they present, which can cause ignition and combustion problems [8]. Decreasing the moisture has positive effects on the flue gas temperature, ignition properties, wall heat flux, flame stability and char burnout [9], but it increase their flammability tendency.

There are three types of co-firing [10]:

- Direct co-firing: biomass and coal are burned in the same furnace. It can be that both of them are handled and milled in advance together or in different systems and mixed just before introducing them into the furnace. This is the most used co-firing type.
- Indirect co-firing: the biomass is firstly converted into a clean fuel gas, which can then be burnt in the same furnace as the coal.
- Parallel co-firing: consists on installing a different boiler for biomass in the steam system of the coal power plant.

6.2 Ignition Characteristics

Conceptually, fuel mixtures are an excellent solution for the current energetic and environmental problem. It is important to remark that biofuels are new materials that need to be much further studied and characterised. Mixing these materials is creating even newer and less known materials, whose properties are unknown and whose risks need to be carefully determined.

Regarding associated risks, the flammability behaviour of mixtures of solids have been mainly studied with an inert sample as one of the parts of the mixtures. Solid inert materials are commonly used to prevent and mitigate dust explosions, in inerting systems or in explosion suppression [11]. There has been many studies on the behaviour of solid inerts in the flammability parameters of dusts, since different properties of the materials cause a different degree of influence in the flammable dusts. Janes et al. [12] reviewed different studies and concluded that the inerting/explosion suppression efficiency depends notably on the concentrations of inert dusts, the kind of mixtures, the particle size distribution, the ratio of the densities of particles and the heat capacity of the solid inertant.

Furthermore, Janes et al. [12] studied the influence of three different inert solids (alumina, Kieselguhr and silica) mixed with eleven industrial organic dusts. They observed that increasing the content of inertants above a given threshold leaded to higher MIE, but below this threshold, the observed behaviours hardly vary. This threshold was also observed for MITc values. However, it was not observed for every inertant in MITl. By adding only a small concentration of alumina, the MITl values sharply increase, but high concentration were requested to raise it significantly.

Danzi et al. [13] observed that the effect of some inertants as limestone is limited. They mixed it with 10% flour and the temperature only rises to 540 °C, observing a very small influence of the inertant's particle size. On the other hand, they observed a strong dependence upon the particle size distribution of the inertant when studying the behaviour of layers, even promoting the ignition of flour samples. They explained this behaviour as the cracking and curving of a layer of flour when it is heated up, which decreases the surface that is in direct contact with the hot surface. When the inertant is added, due to its density and behaviour, it ends up in direct contact with the hot surface and with the flour, promoting its ignition.

The influence of inertants in the ignition of layers has also been studied by Bideau et al. [14] that observed that sometimes concentrations higher than 85% are needed to avoid ignition of metallic dusts.

It has been detected that the particle size distribution of the solid inertants has a big impact on the flammability of the mixtures, a decrease on the particle size causing an increase on the efficiency of the inerting [15].

Addai et al. [16] studied the influence of three inerting materials (ammonium sulphate, magnesium oxide and sand) in the sensitivity of highly flammable dusts, and they observed the permissible range for the inert mixture to minimize the ignition risk lies between 60 and 80%. They also observed that inerts have a different influence on the change of behaviour, being sand the one with less inerting effect on the combustible dust, probably due to its high bulk density and particle size.

Other researchers have also observed that these thresholds commented below are not as strict as they seemed to, and as a consequence, the common recommendation of solid inertants introduction up to 50–80% to eliminate the dust explosion risk [17] should be reconsidered [18].

These works with mixtures between combustible dusts and inertants only point out the difficulties treating this type of materials. These difficulties are increased when studying the mixtures formed by two or more combustible dusts, and even more if at least one of these materials is a biofuel, still unknown and whose properties need to be better defined before considering it safe.

If we look into the ignition properties of biomass-coal mixtures, the available information is almost inexistent. It is well-known that the ignition characteristics of coal, waste and biomass dust have to be carefully determined in each specific situation due to their heterogeneity, but mean values are used to demonstrate the risk that is associated to their handling and storing. This risk will therefore present in any mixture of these materials, and depends on the proportion of each component in the mixture. However, the influence of each one of the substances in the characteristics of the mixtures cannot be determined with a direct relation.

In previous works, we have determined the flammability properties of mixtures of coal, waste and biomass at different concentrations [19]. Coal and waste presented similar values, and any mixtures of the two of them had characteristics situated in the middle of the range that they limit. On the other hand, biomass present much different characteristics, being easier to ignite, and its influence in the mixtures is high, making the mixtures to tend to its behaviour more than to the one of coal or waste.

By comparing the mixture characteristics with the component characteristics through a mass weighted proportional contribution from each sample, it was observed that a theoretical determination of these parameters through the base values of their former components is not accurate and provides more relaxed values than the ones needed on a facility treating this type of samples.

References

1. Mokrzycki E, Uliasz-Bocheńczyk A (2003) Alternative fuels for the cement industry 74:95–100
2. Sahu SG, Chakraborty N, Sarkar P (2014) Coal–biomass co-combustion: an overview 39:575–586
3. EPACN EP-C (2008) Environmental and sustainable technology evaluation-biomass co-firing in industrial boilers–University of Iowa
4. Al-Naiema I, Estillore AD, Mudunkotuwa IA, Grassian VH, Stone EA (2015) Impacts of co-firing biomass on emissions of particulate matter to the atmosphere. Fuel
5. Werther J, Ogada T (1999) Sewage sludge combustion. Prog Energy Combust Sci 25(1):55–116
6. Van Loo S, Koppejan J (2002) Handbook of biomass combustion and co-firing. Prepared by task 32 of the implementing agreement on bioenergy under the auspices of the international agency. Twente University Press, Enschede
7. Hughes E (2000) Biomass cofiring: economics, policy and opportunities. Biomass Bioenerg 19(6):457–465
8. Sami M, Annamalai K, Wooldridge M (2001) Co-firing of coal and biomass fuel blends 27:171–214
9. Tan P, Ma L, Xia J, Fang Q, Zhang C, Chen G (2017) Co-firing sludge in a pulverized coal-fired utility boiler: combustion characteristics and economic impacts. Energy 119:392–399
10. ETIP Bioenergy (2019) Biomass co-firing. http://www.etipbioenergy.eu/value-chains/conversion-technologies/conventional-technologies/biomass-co-firing. Accessed 11 Sep 2019
11. Amyotte PR (2006) Solid inertants and their use in dust explosion prevention and mitigation 19:161–173
12. Janès A, Vignes A, Dufaud O, Carson D (2014) Experimental investigation of the influence of inert solids on ignition sensitivity of organic powders. Process Saf Environ Prot 92(4):311–323
13. Danzi E, Marmo L, Riccio D (2015) Minimum ignition temperature of layer and cloud dust mixtures. J Loss Prev Process Ind 36:326–334
14. Dufaud O, Perrin L, Bideau D, Laurent A (2012) When solids meet solids: a glimpse into dust mixture explosions. J Loss Prev Process Ind 25(5):853–861
15. Amyotte PR, Mintz KJ, Pegg MJ, Sun Y-H, Wilkie KI (1991) Effects of methane admixture, particle size and volatile content on the dolomite inerting requirements of coal dust. J Hazard Mater 27(2):187–203
16. Addai EK, Gabel D, Krause U (2016) Experimental investigations of the minimum ignition energy and the minimum ignition temperature of inert and combustible dust cloud mixtures. J Hazard Mater 307:302–311
17. Eckhoff R (2003) Dust explosions in the process industries: identification, assessment and control of dust hazards. Gulf Professional Publishing
18. Dufaud O, Perrin L, Bideau D, Laurent A (2012) When solids meet solids: a glimpse into dust mixture explosions 25
19. Fernandez-Anez N, Slatter DJF, Saeed MA, Phylaktou HN, Andrews GE, Garcia-Torrent J (2018) Ignition sensitivity of solid fuel mixtures. Fuel 223

Part II
Industrial Position

Chapter 7
Accidents

Accidents in the process industry result into human and material losses, accident cost, production stop cost, judgements, etc. In order to avoid all those consequences, improvement of industrial safety measures must be achieved, but it cannot be accomplished without a deep knowledge of their primary causes, spread processes, etc., so the risk assessment can be carried out properly, and the safety measures defined.

Among all the possible accidents that can occur when handling solid biofuels, fires and explosions are the most common ones but also the most dangerous. Fires may occur due to combustion when external ignition source is applied or due self-ignition process when biofuels are stored in piles or bulks. Quite often, fires may result into explosions thus aggravating the consequences and losses. On the other hand, dust explosions are not only one of the most serious explosion hazards, along with vapour cloud explosions, VCE, and boiling liquid expanding vapour explosions, BLEVE, but also, quite common in the process industry [1].

If the dust is dispersed in the air resulting into a dust cloud and ignites a fireball may occur, if the cloud is not confined, or a dust explosion, if the cloud is confined. On the other hand, if the dust cloud does not ignite, the particles will deposit on the surfaces as a layer whose ignition will produce a fire. Some substances, when deposited in a layer, may decompose and produce some flammable gases whose inflammation might have tremendous consequences, but usually this won't happen regarding biofuels.

Besides self-ignition process, any other fire or explosion will be produced because of an ignition source such as flames, heat, hot surfaces, shock waves, incandescent material, sparks (mechanical, due to friction or impact; electrical, electrostatic, static) and lightening.

7.1 Fires

Three parameters are needed to produce a fire (that could also prompt an explosion): oxidant, fuel and ignition source. The oxidant, also called oxidizing agent, is a substance capable to oxidize other substances which, in the present case, are

N. Fernandez-Anez et al., *Explosion Risk of Solid Biofuels*,
SpringerBriefs in Energy, https://doi.org/10.1007/978-3-030-43933-0_7

the fuels. The ignition source produces the combustion of this mixture generating the first stage of the fire in which the oxygen amount in the environment is around 20.5%, surrounding temperature is approximately 40 °C and the flame temperature reaches 530 °C. In this incipient stage the combustion produces hot ascendant gases.

Second stage would be fire growth in which the heat release rate will increase, due to convection phenomena and radiant heat from the fire and the hot particles. In this stage, more hot gases will be produced, and as the heating increases, they expand producing a pressure increase if confined. The surrounding oxygen will be reduced, and the temperature increased up to 704 °C. When the fire is confined, after the plume reaches the ceiling, it will extend horizontally.

As the stage continues, the governing heat transfer mechanism will be thermal radiation instead of convection, which means that the heat flux will increase, and flashover may happen. Flashover is a near-simultaneous ignition, a sudden combustion of the remaining fuel. Flashover temperature range comes from 500 to 600 °C, and the heat flux varies from 15 to 20 kW/m^2.

This phenomenon will produce a fast transition to the third stage: full developed fire. If the fuel and oxygen amounts is not enough or the heat release rate is not high enough, flashover won't be produced, but the third stage would be achieved in a gradual transition, until the entire fuel is burnt. In this stage, the energy release will reach its maximum, and the gasses will continue to heat reaching temperatures between 700 and 1200 °C.

Once that the entire amount of fuel is burnt or the available oxygen is not enough (it has been mostly consumed), the fourth stage will be achieved: decay stage. In this stage smouldering is produced, and the surrounding temperatures will be above 600 °C and the environment oxygen content will be less than 15%. Significant amounts of carbon monoxide are produced during this stage and, if its temperature and the fuel gas temperature in the smoke is greater than its ignition temperature a blackdraft might happen. In a blackdraft a deflagration is produced when a cool vent interferes inside a confined hot gases environment with low oxygen content (Fig 7.1).

Fire accidents in process industry are mostly due to liquid or gas fuels so pool fires, boil over, fireballs or jet fires are produced. Regarding solid fuels, and more precisely solid biofuels, the main causes for fire accidents is self-ignition process and layer ignition.

In self-ignition process the substance begins its combustion lacking an external ignition source. Because of the self-heating of the material, the combustion is produced if the generated heat is not dissipated into the environment, which means that temperature rises, and the substance oxidizes. This process is more common when the material is stored in bulks or piles, as the exothermic oxidation of the inner material produces amount of heat that cannot be dissipated but heating the surrounding material, which makes it appropriate for biofuels treatment and storage industries.

Layer combustion propagation will occur very slowly if compared to dust cloud combustion propagation, due to the low oxygen ratio. Usually, layer combustion begins with a smouldering process rather than an inflammation and its defined by the thickness of the layer, particle size, material composition and air flow along the layer.

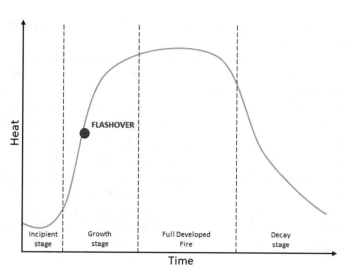

Fig. 7.1 Fire stages

7.1.1 Real Cases of Fires in Biofuels Industries

Merseyside (UK, 2011): Sonae wood chipping factory had faced fatal accidents before. In 2002 a dust explosion produced one serious injury, in 2010 there was a fire that took 4 days to extinguish and the same year two workers of the plant were dragged into the machinery of a storage silo. But it was the vast fire in 2011 (and a second small fire in 2012) the one that caused the final closure of the facility in September 2012.

The fire began the night of the 9th of June 2011 at Sonae facility. This factory was a recycler of waste wood producing wood particle board. The fire was originally produced in the wood chip storage hopper in the chip storage and particle preparation area, in the first of six bunkers, so when the fire propagated and broke the remaining concrete bunkers in which 12,000 tonnes of wood chips were stored, the fire increased and took eight days to extinguish. The fire burnt chemicals located in the facility generating a toxic smoke that affected more than 200 people causing non-permanent eye, skin and respiratory tract injuries.

Copenhagen (Denmark, 2012): On December 17, 2012, a fire broke at Amager CHP plant. The combined heat and power plant was Vattenfall's property and it co-fired coal and biomass. The fire was produced at the wood pellets storage silo and took several days to extinguish it. A few months before this fire accident, in May, an explosion took place at the same plant.

Tilbury (UK, 2012): The Tilbury B Power Station (one of the stations of the Tilbury Power Plant) co-fired oil and biomass, as well as fired coal. In 2012, a fire occurred in the plant when in the fuel storage area, where the fuel (wood pellets) was, burnt. It was estimated that between 4000 and 6000 tonnes burnt, involving more than 120 firefighters. The fire was located at storage hoppers unit 9 and unit 10 at 33 m above ground level. The cause of the fire remains uncertain, however the Fire and Rescue Service of Essex County stated three hypotheses. In the first one the wood pellets dust was in contact with the lighting units and began to smoulder; in the second one, the conveyor that was placed above the hoppers produced the ignition of the wood pellets due to overheat; in the third one the fire was caused because of an electrical failure that produced the ignition of the surface wood pellets of the hoppers.

Tyne (UK, 2013, 2015 and 2019): On October 11, 2013, a fire was produced in a conveying system at Port of Tyne. The conveyor transfer tower was used to transfer wood pellets from the storage facility to the rail loading silo. In order to extinguish the fire of the biomass plant, around 50 firefighters were needed.

Two years later a huge fire was produced when a ship carrying a cargo hold of 11,000 tonnes of wood pellets began to glow because of self-ignition process. The operation involver the discharge of 100 tonnes of pellets in the dock to cool them down.

Six years later, on September 17, 2019, another fire was produced in the Tyne dock when a welding spark produced the ignition of the dust of a biomass pellets storage silo. The firefighters needed 14 h to extinguish the fire cooling the inside of the silo from the outside.

Hexham (UK, 2013): Egger Hexham chipboard factory suffered a huge fire originated in the biomass incinerator caused by the ignition of the thermal oil of the heating system rupturing a pipe. The fire was confined to the heat generating biomass plant that supplies factory's energy.

Yorkshire (UK, 2014): On the 2 June 2014 at R Plevin Recycling Plant (Hazelhead, Sheffield site), during the checking routine, it was discovered a 3000 tonnes wood chip pile smouldering. The pile was isolated, so the fire did not spread but even so, it took more than ten days to extinguish the fire. Because of the accident, the plant was not taking any more wood until the tonnages of biomass on site were considerably reduced. Once that the plant was again operational, the wood stacks were monitored, and the number of wood piles reduced.

Southampton Docks (UK, 2015): The night of January 3, 2015, a fire broke out in a woodchip pile at Southampton Docks. After the fire was extinguished, the wood pile remained smoldering the following days.

Erla (Spain, 2019): On March 2019, a fire took place at Arapellet's facility in Erla (Aragón), Spain, when a biomass pile stored outside the facility burnt. The plant was Forestalia's property and was the bigger wood pellet production plant in Spain. At the moment of the accident, wind was blowing, so the fire propagated inside and outside facility's limits, but it did not cause material or personal losses.

7.2 Explosions

Dust explosions have been considered previously in this book in a theoretical way, so the formation, propagation and mechanisms can be understood. However, in order to be able to define preventive and protective measures, it is important to understand where dust explosions are produced in the process industry.

Dust clouds typical inside process equipment such as mills, mixers, bucket elevators, silos, conveyor belts, dryers, cyclones, etc. That equipment might have steam heating pipes, hot bearings, metal parts rotating and producing sparks, gears that may produce electrical sparks, damaged cables or other electrical pieces, heat emission parts due to friction elements...

All those processes produce fine dust particles and some of that equipment is used in processes that requires air flow or turbulence, so the explosive atmosphere is produced, as for example dryers or mills. In other cases, the dust forms a cloud when moving the material, for example charging and discharging silos, transporting material on conveyors or bucket elevators [2].

7.2.1 Domino Effect

When a primary accident triggers secondary accidents, it is called domino effect. Regarding dust explosions, domino effect happens often, mostly producing fires after blast and, of course, increasing the human and material losses. In dust explosions, the primary explosion is the first one that takes place in a process unit, while the secondary explosion is produces outside que process unit. It is possible to generate secondary explosive clouds by the blast wave from a primary dust explosion entraining layers and deposits of dust [3]. When the second explosive atmosphere is formed, and a primary blast has happened, the chances of having a secondary explosion become high. Domino effect is not limited to two explosions, but it may produce as many explosions as the conditions allow it. According to Lees [4] the secondary explosion will involve greater amounts of combustible dusts, so the consequences might be increased.

Biomass industry requires major hazardous installations (MHIs) working with flammable materials, flammable dusts, high temperature process, and other dangerous conditions, so domino effect becomes probable if the primary explosion is not mitigated. Because the domino effect explosions take place in different units of the facility, preventive measures are difficult to implement as all the process areas need to be considered as one. Domino effect has become an important research topic [5, 6] in the last years as its consideration and understanding is the main tool to implement prevention and mitigation measures.

7.2.2 Real Cases of Explosions in Biofuels Industries

Hebei (China, 2010): On February 2010, a dust explosion took place in Qinhuang-dao, in the province of Hebei, inside a starch factory. The explosion took 19 lives, and 49 people were injured as, at the moment of the explosion, more than a hundred people were working in the factory. The starch powder present in the factory ignited producing the dust explosion.

Nijmegen (Netherlands, 2012): The morning of November 8, 2012, an explosion occurred at the power plant of Nijmegen (Gelderland), Electrabel's property (GDF Suez's subsidiary). The facility was a coal and biomass co-firing power plant of 600 MW. The explosion was produced inside one of the boilers, due to a steam pipe overpreassure.

Copenhaguen (Denmark, 2012): On May 2012, an explosion occurred at Amager Power Station. The power station was originally a 70 MW coal power plant, that turned into biomass (wood pellets) power plant on 2010. During the demolition operation of an old silo, that had a bunch of pellets stuck. When trying to release the pellets the explosion was produced, generating a fire inside the wood pellet silo. The technique used to release the pellets, denominated "bang and clean", consisted of small explosions of oxygen and methane to unblock the pellets. Three people were injured in this accident.

Minnesota (USA, 2013): On 2013, at Koda Energy combined heat and power plant, an explosion (that lead a fire) took place. The plant was feed with woody biomass, stored in silos. On April 25 the airborne dust ignited producing the explosion and subsequent fire of two silos storing wood chips and oat shells. The explosion was produced at the top of one of the silos, and transported to the second one by the conveyor. The accident did not cause any personal injury, but both silos were demolished and rebuilt.

Bosley (UK, 2015): The Cheshire wood mill that produced wood flour, exploded on July 2015; causing the dead of four people. The explosion was followed by a huge blast leading to 35 casualties. It took weeks to shut completely down the subsequent fire. The weeks before the explosion, complaints about sawdust were sent to the council enforcement officers who attended the mill. The company (Wood Treatment Ltd) and its director, were charged with corporate manslaughter by gross negligence.

Alberta (Canada, 2019): Pinnacle Renewalble Energy's wood pellet plant in Entwistle, Alberta, suffered an explosion on February 2019. The explosion was produced at one of the dryer units and a subsequent fire damaged the dryer area. The explosion injured twelve workers, most of them not serious injured.

7.3 Poisoning

As biomass is composed by organic matter, it decomposes among time emitting gasses, especially when biomass is compacted in pellets. Regarding pellets, it has been observed that the longer they are stored the greater is the amount of emitted gasses [7] posing a great risk when transporting and storing biomass pellets. The main gasses emitted are carbon monoxide, carbon dioxide and methane, but also volatile organic compounds (VOCs) are also emitted; and the emissions are related to the temperature of the storage room: high temperatures produce more emissions.

As those emissions may cause poisoning, the off-gassing phenomena has been widely studied [9], but the chemical reactions that take place and produce the emissions are not deep understood yet. The most studied case is the lignocelullosic biomass, as some studies have noted that the carbon monoxide off-gassing is mainly produced by the cracking of carbonyl and carboxyl from hemicellulose [8].

VOCs, methane or carbon dioxide do not pose a risk as severe as carbon monoxide emissions. CO produces oxygen depletion in storage facilities, which is difficult to notice, so poisoning may happen, and fatal consequences may occur; and are quite common during marine transportation.

7.3.1 Real Cases of Poisoning Accidents in Biofuels

Saga Spray (Sweeden, 2006): On November 2006, an accident happened in Helsinborg port, in Sweeden, onboard the Saga Spray ship registered from Hong Kong, that was discharging wood pellets from Brisith Columbia (Canada). The ship loaded wood pellets in Canada on September 26, and the accident took place on November 16, so the pellets were stored for 83 days off-gassing. While discharging the pellets, the crane could not reach the pellets located at edges of the cargo, so a bulldozer was used to pile the remaining pellets. In order to do that (which is a normal practice), the bulldozer operator and a seaman entered the cargo through an enclosed stair. At the bottom of the stair, the seaman collapsed and fainted and shortly after also the bulldozer operator collapsed. An operator went in for helping but felt breathing problem, so he came back to the deck and the emergency services were alerted. When they realized that the air was unbreathable, three members of the crew went down to assist the fainted operators using self-contained breathing masks. They retrieve the bulldozer operator but couldn't recover the seaman. When the emergency services arrived, they rescued the seaman, and sent both to the hospital. Unfortunately, the seaman did not survive, and the bulldozer operator was seriously injured. Swedish authorities carried out atmospheric measurements of the cargos after the accident, and at the time of opening the hatch the oxygen percentage was 15%, and the carbon monoxide was over 1000 ppm, while normal atmospheric conditions are 21% of oxygen and about 9 ppm of carbon monoxide.

Amirante (Denmark, 2009): On 15 July 2009, an accident occurred in Amirante cargo ship, when it was transporting wood pellets in bulk from Riga (Latvia) to Copenhagen port (Denmark) for the Amagerværket power plant. During the trip, it was noticed that two seamen were disappeared, and they were found dead at the bottom of the stairwell that connected the deck with the forepeak compartment (stowage compartment at the ship's bow) where buckets, surplus wooden planks, mooring ropes, etc., were stored. When the master of tried to rescue them, he began feeling difficult breathing, so he realized the atmosphere might be poisoned. He tried to wake up the seamen and, as he couldn't, ran back to the deck and ask for medical assistance to the Danish authorities. By the time the doctor arrived, both seamen were dead. The forepeak compartment was connected to the cargo hold through a door that was found not well closed, so the gasses emitted from the wood pellet cargo were travelling to the stairwell. The autopsy confirmed both men died because of carbon monoxide poisoning.

Corina (Denmark, 2015): Corina was a cargo vessel from Poland, having a fatal accident in 2015 when transporting wood pellets from the port of Arkhangelsk (in Russia) to Hanstholm (Denmark). During their trip, a seaman died at a room at a lower deck. When the ambulance arrived, six people were in the accident site and the air was confirmed to be normal for three times. Only after the third time the ventilation was switched on and the paramedic evacuated everybody from the room. Two crew members were feeling so dizzy they couldn't climb the ladder to the exit and remained there until they fainted. Only fire fighters were able to rescue the crew members and retrieve the dead seaman whose autopsy confirmed dead by carbon monoxide poisoning. Four crew members were sent to the hospital and recovered after some weeks.

References

1. Abbasi T, Abbasi SA (2007) Dust explosions-Cases, causes, consequences, and control. J Hazard Mater 140(1–2):7–44
2. Yuan Z, Khakzad N, Khan F, Amyotte P (2015) Dust explosions: a threat to the process industries. Process Saf Environ Prot 98:57–71
3. Yuan Z, Khakzad N, Khan F, Amyotte P (2016) Domino effect analysis of dust explosions using Bayesian networks. Process Saf Environ Prot 100:108–116
4. Mannan S (2005) Lee's loss prevention in the process industries
5. Khakzad N, Khan F, Amyotte P, Cozzani V (2014) Risk management of domino effects considering dynamic consequence analysis. Risk Anal 34(6):1128–1138
6. Cozzani V, Antonioni G, Landucci G, Tugnoli A, Bonvicini S, Spadoni G (2014) Quantitative assessment of domino and NaTech scenarios in complex industrial areas. J Loss Prev Process Ind 28:10–22
7. Arshadi M, Gref R (2005) Emission of volatile organic compounds from softwood pellets during storage. For Prod J 55(12):132–135
8. Rahman MA, Rossner A, Hopke PK (2018) Carbon monoxide off-gassing from bags of wood pellets. Ann Work Expo Health 62(2):248–252
9. Sedlmayer I, Arshadi M, Haslinger W, Hofbauer H, Larsson I, Lönnermark A, Nilsson C, Pollex A, Schmidl C, Stelte W, Wopienka E (2018) Determination of off-gassing and self-heating potential of wood pellets–Method comparison and correlation analysis. Fuel 234:894-903.

Chapter 8
Legislation

Nowadays, environmental issues are the key of policy, economic and social movements. Due to its importance, the world is developing solutions capable to assure a sustainable future. By means of large agreements between countries, a new series of directives and legislation are created allowing to control and reduce the most harmful activities and to boost new clean energy systems.

The policies regarding climate change are leading the industry activity by means of legislations that controls the harmful activities and limits its impacts. This fact has implied deep changes in the energy sector resulting necessary to drive the activity of the companies to new clean resources.

Solid biofuels can help to solve the environmental problems, therefore, organizations as European Union has encouraged its use by means of legislation. However, biomass handling implies safety problems regarding the formation of explosive atmospheres, as it has been explained before. Therefore, it is also important to analyse the industrial safety legislation to create a clear understanding of current knowledge.

Consequently, the aim of this chapter is to collect and review the current legislation in order to understand the policies that boost biofuels and the safety legislation regarding explosion risk of solid biofuels from an industrial point of view.

8.1 European Directives

Environmental policy of European Union relies on *"the principles of precaution, prevention and rectifying pollution at source"* [1]. Articles 11 and 191 to 193 of the Treaty on the Functioning of the European Union (TFEU) [2] lay the foundation of its environmental policies. EU develops environmental action programmes in all areas of environment policy. They are evolved according to international environmental agreements like Kyoto and Paris agreements. Furthermore, these days result indispensable integrating environmental concerns into other areas as for example, energy. Lately, such integration has made significant developments, being reflected in the long-term programmes as the low-carbon economy by 2050.

N. Fernandez-Anez et al., *Explosion Risk of Solid Biofuels*,
SpringerBriefs in Energy, https://doi.org/10.1007/978-3-030-43933-0_8

EU environmental policy directly affects the industry, driving its changes through environmental action programs by means of its implementation, enforcement, and monitoring of these actions.

One of the most important EU plans is to reach the targets fixed in the Paris Agreement [3]. The European Union leads it under the United Nations Framework Convention on Climate Change agreement. The agreement intends to minimize the climate change and the main point is to reduce the greenhouse gas emissions thought the participation of all the countries. Particularly, the EU goal is to reduce it by at least 60% below 2010 levels by 2050. As over mentioned, this objective has a great impact on industry, forcing it to evolve fast to newer and cleaner technologies. These changes imply a high cost for companies therefore EU also give economical support to companies by means of its granted programmes.

In this sense, to achieve the fixed objectives the EU try to boost new energy sources by means of legislation. The Communication of the Commission to the Council and the European Parliament: A policy framework for climate and energy in the period from 2020 to 2030 try to assure the achievement of the EU targets [4].

The over mentioned communication specifies actions as a 40% reduction below the 1990 level in EU greenhouse gas emissions by 2030 or increasing the weight of renewable energy consumed in Europe to at least 27%. These actions are supported by de Directives shown in Fig. 8.1.

These directives act in the main three environmental issues, energy efficiency [5], renewable energies, and greenhouse gas emission reduction [6]. Regarding solid biofuels boosting, Directive 2009/28/EC of the European Parliament and of the Council of 23 April 2009 on the promotion of the use of energy from renewable sources [7], encouraging the use of biomass as an alternative source of energy being also supported by the Communication from the Commission—Biomass action plan [8]. EU presents biomass as a potential help to minimize climate change since it uses lower greenhouse emissions. Its consumption as an electricity generator, heating fuel and biofuel production can diversify the European Union energy system and can create growth and new jobs.

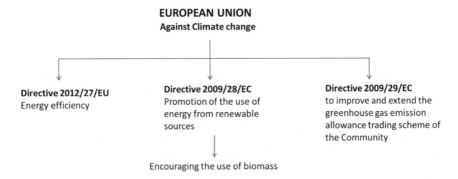

Fig. 8.1 The main European directives to fight against climate change

Under the application of these Directives, the interest of the European Union on biofuel uses seems clear. However, the explosion risk of solid biofuels still poses safety problems in the industry when handled, as explained in Part I of this book.

Safety policies have evolved during the last 40 years, contributing to the current Unique Market. Regarding safety, Directives can be classified in two main groups.

- Social Directives: These Directives regulate the protection of the health and safety of workers.
- New Approach directives: These Directives are referred to products and its aim to create common legislations to ensure free movement of products on the EU market.

On one side, according to the first group, Article 137 of the Treaty specify that the Council may adopt, through the Directives, the minimum requirements in the working environment, to guarantee a better level of protection of the health and safety of workers are regulated through of the Directive 1999/92/EC of the European Parliament and of the Council of 16 December 1999 on minimum requirements for improving the safety and health protection of workers potentially at risk from explosive atmospheres [9]. This directive establishes the mode of operation within the installation, as well as the adequacy of the equipment to the work areas defined under this directive, affecting the responsibility of the owner of the installation. The Directive also lays down the employer liability as a last responsible for accidents and incidents that may occur in an industrial environment. The employer must apply the necessary preventive measures. This directive is also called ATEX 137.

On the other side and according to the second group, Directive 2014/34/EU of the European Parliament and of the Council of 26 February 2014 on the harmonization of the laws of the Member States relating to equipment and protective systems intended for use in potentially explosive atmospheres is referred to the products safety, and to the equipment and products for use in explosive atmospheres [10]. It establishes the Essential Safety Requirements that these products must meet, as well as the procedures for conformity assessment. Therefore, this directive, called ATEX 95, establishes the responsibility of the manufacturers of such equipment. It is based on Article 95 of the EC Treaty then, the directive is aimed at manufacturers and designers. Article 95 ensure that industrial products placed in marked comply the essential requirements. The main differences between ATEX 137 and ATEX 95 are shown in Fig. 8.2.

Suppliers, manufacturers or importers must ensure that their products meet essential safety and health requirements under adequate conformity procedures. Therefore, the products must be tested and certificated by a 'third-party' certification body, also called Notified Body, or by 'self-certify' depending on the zone classification where the equipment is intended to use. Once the equipment is certified, it is marked by the 'EX' symbol to identify it as an appropriate equipment intended for use in explosive atmospheres.

Fig. 8.2 ATEX directives
and its marking

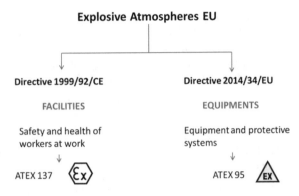

8.2   National Regulations

European countries must follow the environmental and safety Directives as well as the companies who want to make activities in UE or place products in the European market. However, each country has its own regulation that sometimes, can be more restrictive than directives. Furthermore, countries as United States has different legislation and different points of view about how to reach a sustainable future.

For that reason, it is essential for companies and industry to know these differences in order to take advance of the favourable policies or to prevent problems to overcome due to the lack of incentives. Hence, this section collects the National Regulation regarding the use of solid biofuels and its safety issues of Spain, Germany, United Kingdom and United States.

8.2.1   Environmental and Energy Regulations

The grand worldwide agreements regarding environmental issues are the responsible of policy trend all over the world. Nevertheless, each country has its own legislation to boost the needed changes. Nowadays, the support of the countries in the development of the renewable energies plays an essential role in industry since new technologies imply higher investments than conventional technologies. In fact, some renewable energies depend on governmental incentives to compete with conventional energy systems.

Countries belonging European Union have been supported by programmes as Horizon 2020 and will be able to participate in next framework, Horizon Europe. These programmes finance projects allowing to generate the bases and the know-how for industry, to help them to be more competitive in the main industrial sectors, such as the energy sector. In the case of United States, the Office of Energy Efficiency and Renewable Energy offers support to ensure that US has a broad and economical supply of clean energy.

Table 8.1 Resume of the main laws and policies of Spain, UK, Germany and US regarding clean energy and environment

	Spain	UK	Germany	US
Environmental legislation	Royal Law-Decree 34/2007 Royal Decree 1/2005	Climate Change Act 2008 (Order 2019)	Measures to fight against climate change	CAA Clean Air Act National Ambient Air Quality Standards (NAAQS)
Clean energy legislation	Royal Decree 661/2007	RO-Renewable legislation Energy Act 2013	EEG—The renewable Energy Sources Act	Energy policy act US

A part of the support of these organizations, the Table 8.1 shows the specific legislation of 4 countries regarding Environmental and Energy issues.

1. **Spain**

In Spain, the air quality and the greenhouse emissions are controlled by the Royal Law-Decree 34/2007 [11] and the Royal Decree 1/2005 [12] respectively. The first one states that the Competent authorities will regularly evaluate the air quality in their corresponding territorial area, and that the Government, with the participation of the autonomous communities, may establish by Royal Decree the emission limit values for pollutants. In addition, it catalogues potentially contaminated activities. The second one regulates the greenhouse gas emission allowance trading regime. It transposes Directive 2009/29/EC, establishing a regime for the trading of greenhouse gas emission rights in order to fulfill the Kyoto commitments [13]. From 2021 to 2030, Spain intend to reduce its greenhouse emissions a 40% comparing to the emissions in 1991.

Regarding energy policies in Spain, at a national level, the most important legislation concerning the production of electric energy through renewable energy sources is Royal Decree 661/2007 [14], which regulates the activity of electric power production under a special regime. The R.D. creates a favorable scenario for biomass cogenerations, especially in the tertiary sector.

2. **United Kingdom**

In the case of United Kingdom, the Climate Change Act of 2008 [15] stipulated a greenhouse gas emissions reduction target in 2005 from at least 80% to at least to 100%. UK intends to reach the 'net zero' that would constitute the end of the UK's contribution to climate change. The application of this law will make this country pioneer in the area of climate change.

Regarding the boost of biomass use in UK, Renewable Energy Strategy [16] states that biomass for heat and power has the potential to meet will help to achieve the goal set. The main support for biomass in UK has been directed at electricity generation through policies such as the RO [17]. According to RO, biomass is considered as a renewable energy source if at least 90% of its energy content comes from biomass. A few years ago, one ROC would correspond to one MWh of eligible renewable electricity generation independently of renewable energy resource, however, in 2009 the renewable energy technologies started to be banded into different categories. This measure allowed to boost the clean technologies in low states of development, while the most established technologies received less ROC support.

3. Germany

Conversely, as happens in Spain, in Germany, the set target to reduce the gases emissions in 2050 is only a 50% less than the total amount of emissions in 1990. Even so, the use of biomass is also boost through the Renewable Energy Sources Act or EEG [18]. Under the application of this regulation, the consume of biomass as an energy source has changed significantly. This legislation consists of a series of laws whit the objective to create and promote a scheme to encourage the generation of electricity from renewable sources. The EEG has amended during the last years in order to follow the global trend in terms of gas emission reduction, but even after the last actualization in 2019, the set targets are significantly inferior than other European countries. The Directive 2009/28/EC requires Germany to produce in 2020 at least 18% of its gross final energy consumption, considering transport and heat from renewable energy sources.

4. United States

Finally, In US, the policies have been slightly out of alignment of the great environmental agreement as Kyoto or Paris. Moreover, US is out of the Directive 2009/28/EC so its legislation have notable differences comparing the overmentioned countries. Clean Air Act (CAA) [19], gave the Environmental Protection Agency (EPA) the capability to act fighting climate change. The CAA establishes that the states regulate sources of air pollution with specific requirements for industrial entities to address and control their contributions to air pollution. Nevertheless, under the application of this law, nowadays some pollution levels remained above the grant agreements limits in certain parts of the United States as the regulation change depending on the state.

Regarding Energy policy, the Energy Policy Act [20] described measures and policies to fight energy problems, driving US energy policy by providing tax incentives for new clean energy production. Nowadays, according to the Annual Energy Outlook 2019 [21] and depending on the State, the set target of clean energy use varies from 15 to 100% [22].

Concerning biomass use in US, a part of the overmentioned law, the federated law 40 CFR part 80 [22] regulate its use.

The use of biomass as a clean energy resource has regularly been presented in US legislation as a potentially alternative to reduce the foreign oil dependence, and rural economic development, and as a tool to improve the environment problems in the country. Efforts to promote the use of biomass for power generation have focused on wood, wood residues, and milling waste primarily.

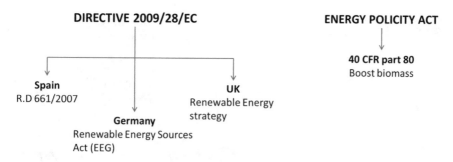

Fig. 8.3 Resume of the main energy policies of Spain, UK, Germany and US

The resume of the different national legislations collected in this chapter are exposed in Fig. 8.3.

8.2.2 Safety Regulations

Besides the safety problems regarding the biomass use in industry, the European Union is regulated by the ATEX directives. The ATEX scheme is needed for industrial activities in Europe and for placing in the market ATEX equipment. Nevertheless, these regulations cannot be used globally, consequently, countries like US need to use an international scheme. IECEx, International Certification for Explosive Atmosphere Equipment born as a system to facilitate the trade of equipment and services for use in explosive atmospheres at an international level and to ensure that this equipment and services maintain an adequate level of safety. This scheme intends to become an international recognized test and certification procedure for explosion protection in industry. Nowadays it is used in more than thirty countries, including, United Kingdom, Spain, Germany, and US.

At technical level, ATEX and IECEx are quite similar. The standards used to meet ATEX are normally European standards as the IEC 60079 [23] series. Compared with IECEx, ATEX directive defines equipment groups and categories, giving ATEX equipment additional label markings. Figure 8.4 summarizes its use in the different countries.

According to European Union countries like Spain, Germany or UK, the requirements of the ATEX directive are transposed to the right of each EU Member by its national regulations.

1. **Spain**

In Spain the Directive 1999/92/CE is transposed through Royal Decree 681/2003 [24]. The preventive measures that are imposed in the industry by current legislation include the following (Fig 8.5):

Fig. 8.4 Applicability of
ATEX and IECEx schemes

Fig. 8.5 Transposed laws in
Spain regarding explosive
atmospheres Directives

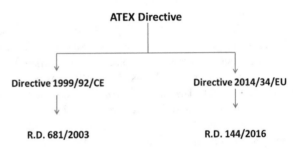

- Characterization of solid substances that may occur in powder form, to assess their risk of explosion and the severity of the explosion.
- Zones classification regarding the probability of the formation of a dust cloud.
- Design of process equipment to prevent the formation of dust clouds.
- Selection and use of electrical equipment without any energy manifestation (including non-electrical equipment), designed to prevent them from being the origin of the initiation of the explosion.
- Selection and use of devices able to prevent the transmission of the explosion or to limit its effects.
- Design of suitable geometric conditions to confine possible explosions and prevent their spread.
- Use of explosion detection devices, which trigger ultrafast extinguishing systems .

All these measures must be included in the Explosion Protection Document, which the employer must compile and keep up to date. Moreover, Directive 2014/34/EU is transposed in Royal Decree 144/2016 [25] which stablish the essential safety requirements for equipment and systems of protection for its use in potentially explosive atmospheres.

2. **United Kingdom**

In UK the requirements of Directive 1999/92/EC were put into effect through regulations 7 and 11 of the Dangerous Substances and Explosive Atmospheres Regulations (DSEAR) [26] (Fig 8.6). DSEAR require employers to control the risks related to fire and explosions. The employers must:

- Evaluate and locate the dangerous substances present in the workplace and analyze its risks.
- Develop measures to remove those risks or, where not possible, control them.
- Put controls in place to reduce the effects of any incidents involving dangerous substances.
- Prepare plans and procedures to deal with the serious and fatal accidents, or incidents and emergencies.
- Train employees about the risks or how to deal with them in the work place, related to dangerous substances.
- Classify and identify installation zones where explosive atmospheres can occur and avoid ignition sources.

As the Directive establish, DSEAR demand a classification into hazardous zones based on the risk of an explosion occurring and protected from sources of ignition by selecting adequate equipment.

The equipment requirements are regulated through Directive 2014/34/EC and are put into effect through BIS Equipment and Protective Systems Intended for Use in Potentially Explosive Atmospheres Regulations [27]. The Regulations apply to all equipment intended for use in explosive atmospheres, whether electrical or mechanical, and to protective systems .

3. **Germany**

In Germany, explosion protection has the same structure, and the ATEX Directives are its base: Directive 2014/34/EU and 1999/92/EC. Consequently, the national laws implement the main points of the directives as follows:

- Regulation on the explosion protection.
- Regulation on hazardous substances.

The regulations propose an explosion protection concept where employers must follow these measures: Prevent the formation and limit the explosive mixtures occurrence by means of inertisation, in the primary explosion limiting their concentrations by increasing air exchange and in the secondary explosion to avoid all the effective ignition sources.

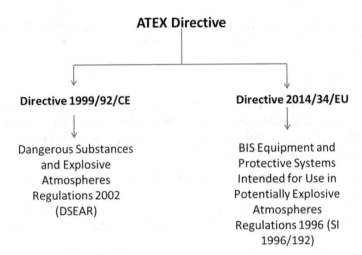

Fig. 8.6 Transposed laws in UK regarding explosive atmospheres

4. **United States**

Finally, in US there are safety and healthy local regulations in use while IECEx has been widely adopted as well. National Electrical Code (NEC) [28], regulate the minimum standards for the safe installation of electrical wiring and equipment in the US.

NEC lays down the responsibility of the employer as the responsible to ensure the application of the correct standards, use the appropriate equipment and to have a Hazard Plan in workplace. In the United States, as the EU Directives also stablish, the equipment must meet the NEC/NFPA-70 requirements. Moreover, equipment can be approved, depending on its constructive requirements, for divisions or zones.

In the case of United States, the Occupational Safety and Health Administration (OSHA) accredits Nationally Recognized Testing Laboratories (NRTL). This organization test and certify the equipment and installations according US standards. Once the equipment is certified, as ATEX and IECEx standards, it is put on the market with the mark of the certifying body.

8.3   Standards

Solid biofuel industry must assure a safety work environment as some biomass dust result highly flammable. This dust is produced and accumulated during the processes generating potential explosive atmospheres. The associated risk to handle flammable dust implies that solid biofuel industry works under ATEX directive, then this section intends to collect the most important standards related to safety in explosive atmospheres. The application of this standards is essential to trade whit the European Union in the case of ATEX Directive but also to ensure a safety work environment in other countries around the word regarding IECEx scheme.

As over mentioned, the standards used to meet ATEX requirements are the IEC 60079 series of standards.

The first group correspond to the series of standards based on ATEX Directive 2014/34/EU. The main standards regarding dust protective systems according this directive are listed below.

- IEC 60079-0:2017: Explosive atmospheres—Part 0: Equipment—General requirements [23].
- IEC 60079-11:2011: Explosive atmospheres—Part 11: Equipment protection by intrinsic safety "i" [29].
- IEC 60079-18:2014: Explosive atmospheres—Part 18: Equipment protection by encapsulation "m" [30].
- IEC 60079-31:2013: Explosive atmospheres—Part 31: Equipment dust ignition protection by enclosure "t" [31].

The second group corresponds with the series of standards based in 1999/92/EC Directive. The Directive stablishes the requirements regarding installations in both non-hazardous and hazardous locations when it is possible the generation of an explosive atmosphere. The main standards of this group are listed below.

- EN/IEC 60079-14:2016: Explosive atmospheres—Part 14: Electrical installations design, selection and erection [32].
- EN/IEC 60079-17:2014: Explosive atmospheres—Part 17: Electrical installations inspection and maintenance [33].

Additionally, to have a better knowledge about the dust flammability it is essential to carry out a dust characterization. To characterize the samples, certain tests are performed so it is possible to determine flammability characteristic as MIT, MIE, LOC, etc., as explained in Chap. 4 of this book. The zone classification, risk assessment, and the prevention and protection measures can be carried out departing of the results. In 2016, the test for the determination of the flammability parameters were collected in the standard EN ISO/IEC 80079-20-2 [34]. Nevertheless, some of these parameters can be determined also following the next standards:

- EN 14034-1:2005+A1:2011: Determination of explosion characteristics of dust clouds—Part 1: Determination of the maximum explosion pressure p_{max} of dust clouds [35].
- EN 14034-2:2006+A1:2011: Determination of explosion characteristics of dust clouds—Part 2: Determination of the maximum rate of explosion pressure rise $(dp/dt)_{max}$ of dust clouds [36].
- EN 14034-3:2006+A1:2011: Determination of explosion characteristics of dust clouds—Part 3: Determination of the lower explosion limit LEL of dust clouds [37].

Table 8.2 Combustible dust classification according to TDG Model Regulations

Class 4	Division 4.1 Flammable solids, self-reactive substances and solid desensitized explosives
	Division 4.2 Substances liable to spontaneous combustion.
Class 5	Division 5.1 Oxidizing substances

- EN 14034-4:2005+A1:2011: Determination of explosion characteristics of dust clouds—Part 4: Determination of the limiting oxygen concentration LOC of dust clouds [38].

The handling of solid biofuels is not only affected by ATEX legislation as some solid biofuels can react and generate accidents during its transportation, storage and, as over mentioned, at the workplace.

The recommendations on the Transport of Dangerous Goods (TDG) contains the needed criteria and methods to guarantee the safety in the transport of dangerous goods. The TDG consist of two different parts, the Model Regulations, volume I and II, and the Manual of test and Criteria.

The aim of the Model Regulations [39] is to state the definition and classification of classes, listing of general packing requirements, the principal dangerous goods, marking, testing procedures, labelling or placarding, and defining transport documents. Regarding solid biofuels, its classification can be resumed as Table 8.2 shows.

The Manual of Test and Criteria [40] allow the classification of dangerous goods by means of technical criteria, test methods and procedures according to the provisions of the Model Regulations. By means of the application of the Manual of Test and Criteria the solid biofuels can be classified as exposed below.

- Section 33. Classification procedures, test methods and criteria relation to Class 4.

  - Division 4.1.

    Test N.1. Test method for readily combustible solids.
    Test N.2. Test method for pyrophoric solids.

  - Division 4.2.

    Test N.4. Test method for self-heating substances.

- Section 34. Classification procedures, tests methods and criteria relating to oxidizing substances of Division 5.1.

  - Division 5.1.

    Test O.1. Test of oxidizing solids.

References

1. European Parliament (2019) Environment policy: general principles and basic framework. EU fact sheets. European parliament. Fact sheets on the European Union European parliament
2. European Union (2012) Consolidated version of the treaty on the functioning of the European Union, vol 47, no 6, pp 47–390
3. European Commission (2015) The paris protocol-a blueprint for tackling change beyond 2020, vol 17, pp 1–17
4. European Commission (2014) COM(2014) 15 final: a policy framework for climate and energy in the period from 2020 to 2030, pp 1–18
5. European Parliament (Oct 2012) Directive 2012/27/EU of the European parliament and of the council of 25 October 2012 on energy efficiency, pp 1–56
6. European Parliament (2009) Directive 2009/29/EC of the European parliament and of the council of 23 April 2009 amending. Directive 2003/87/EC so as to improve and extend the greenhouse gas emission allowance trading scheme of the Community, vol 140, pp 63–87
7. European Parliament (2009) Directive 2009/28/EC of the European parliament and of the council of 23 April 2009 on the promotion of the use of energy from renewable sources and amending and subsequently repealing. Directives 2001/77/EC and 2003/30/EC, vol 1, pp 32–38
8. European Parliament (2005) Communication from the commission: biomass action plan, pp 1–47
9. European Parliament and European Council (2000) Directive no 1999/92/EC, of 16 December 1999, on minimum requirements for improving the safety and health protection of workers potentially at risk from explosive atmospheres, no L 23, pp 57–64
10. European Commission (2014) Directive 2014/34/EU of the European parliament and of the council of 26 February 2014 on the harmonisation of the laws of the member states relating to equipment and protective systems intended for use in potentially explosive atmospheres (recast), pp 309–356
11. Spain (2007) Ley 34/2007, de 15 de Noviembre, de calidad del aire y protección de la atmósfera. Boletín of. del Estado, vol BOE-A-2007, no 16 Noviembre de 2007 (275), pp 1–15
12. Spain (2005) LEY 1/2005, de 9 de marzo, por la que se regula el régimen del comercio de derechos de emisión de gases de efecto invernadero. Boletín Oficial del Estado Español, pp 1–47
13. United Nations (1998) Kyoto protocol to the united nations framework convention on climate change united
14. Ministerio de Industria (2007) Real Decreto 661/2007, de 25 de mayo, por el que se regula la actividad de producción de energía eléctrica en régimen especial, vol 22846–2288, pp 1–41
15. UK Government (2019) The climate change act 2008 (2050 Target Amendment) order 2019, vol 2, no 6, pp 1–2
16. Chaubey R, Sahu S, James OO, Maity S (2013) A review on development of industrial processes and emerging techniques for production of hydrogen from renewable and sustainable sources. Renew Sustain Energy Rev 23:443–462
17. Secretary of State UK (2015) The renewables obligation order 2015 (RO), pp 1–100
18. German Government (July 2017) Renewable energy sources act. EEG, pp 1–179
19. US Goverments Information GPO (2019) Clean air act, pp 1–455
20. US Goverments Information GPO (2005) Energy policy act of 2005, vol 41, pp 1–511
21. US Energy Information Administration (2019) Annual energy outlook 2019 with projections to 2050
22. Enviromental Protection Agency (2015) 40 CFR Part 80. Renewable fuel standard program: standards for 2017 and biomass-based diesel volume for 2018, vol 80, no 239, pp 34778–34816
23. European Committee for Standardization CEN-CENELEC, EN-60079-0:2013 (2013) Explosive atmospheres. Part 0: equipment. General requirements
24. Ministerio de la Presidencia (2003) Real Decreto 681/2003 de 12 de junio, sobre la protección de la salud y la seguridad de los trabajadores expuestos a los riesgos derivados de atmósferas explosivas en el lugar de trabajo, pp 23341–23345

25. Ministerio de Industria y Turismo (2016) Real Decreto 144/2016, de 8 de Abril, por el que se establecen los requisitos esenciales de salud y seguridad exigibles a los aparatos y sistemas de protección para su uso en atmósferas potencialmente explosivas y por el que se modifica el Real Decreto, pp 1–40
26. Health and Safety Executive (2013) Dangerous substances and explosive atmospheres regulations 2002 (L138), vol 138, pp 9–120
27. Department of Trade and Industry, Great Britain (2016) The equipment and protective systems intended for use in potentially explosive atmospheres regulations 1996, no 1107, p 38
28. National Electrical Code Committee (2017) National Electrical Code (NEC) NFPA 70, pp 1–881
29. International Electrotechnical Commission, IEC 60079-11 (2011) Explosive atmospheres–Part 11: equipment protection by intrinsic safety "i"
30. AENOR, UNE-EN 60079-18:2016/A1:2018 (2018) Explosive atmospheres-Part 18: equipment protection by encapsulation "m"
31. AENOR, UNE-EN 60079-31:2016 (2016) Explosive atmospheres-Part 31: equipment dust ignition protection by enclosure "t"
32. AENOR, UNE-EN 60079-14:2016 (2016) Explosive atmospheres-Part 14: electrical installations design, selection and erection
33. AENOR, UNE-EN 60079-17:2014 (2014) Explosive atmospheres-Part 17: electrical installations inspection and maintenance
34. European committee for standardization CEN-CENELEC, UNE-EN ISO/IEC 80079-20-2:2016/AC:2017 (2017) Explosive atmospheres-Part 20–2: material characteristics-combustible dusts test methods
35. AENOR, UNE-EN 14034-1:2005+A1:2011 (2011) Determination of explosion characteristics of dust clouds-Part 1: determination of the maximum explosion pressure p_{max} of dust clouds
36. AENOR, UNE-EN 14034-2:2006+A1:2011 (2011) Determination of explosion characteristics of dust clouds-Part 2: determination of the maximum rate of explosion pressure rise $(dp/dt)_{max}$ of dust clouds
37. AENOR, UNE-EN 14034-3:2006+A1:2011 (2011) Determination of explosion characteristics of dust clouds-Part 3: determination of the lower explosion limit LEL of dust clouds
38. AENOR, UNE-EN 14034-4:2005+A1:2011 (2011) Determination of explosion characteristics of dust clouds-Part 4: determination of the limiting oxygen concentration LOC of dust clouds
39. United Nations (2019) Recommendations on the transport of dangerous goods. Model regulations, vol I, p 470
40. United Nations (2017) Recommendations on the transport of dangerous goods. Manual of tests and criteria, p 38

Chapter 9
Safety Measures

The flammable nature of solid biofuels has been proved, and also the risky conditions on which those biofuels are produced, treated, stored or used. Because of that, it is necessary to stablish safety measures in order to avoid accidents by reducing risks. Those safety measures can be divided in two categories: prevention and protection measures. The first ones—prevention measures—intend to avoid the formation of the conditions that may lead to an accident. On the other hand, protection measures intend to reduce the consequences and effects of the formation of those conditions. In other words, while prevention measures avoid the formation of explosive atmospheres, protection measures minimize the effects of that explosive atmosphere.

It means that prevention measures work on the proper atmosphere and the ignition sources and protection measures work on isolating systems, suppressing the explosion, resistance and venting, etc., so the explosion is suppressed or developed in controlled and safe conditions.

Safety measures should be implemented at the design stage of the facility, because it is the easiest and most efficient way to apply them. Nevertheless, also when the facility is working, those measures can be implemented or upgraded, so safety requirements are achieved. In order to define the necessary measures risk assessment must be carried out so it is possible to identify high risk operations and areas.

9.1 Prevention

As it has been said, prevention measures intend to avoid an explosion or a fire, so it will act on the two parameters that lead to an accident: explosive atmospheres and ignition sources. In other words, prevention acts on the three factors of the combustion triangle: oxygen, fuel and ignition. Analysing the triangle, the main and elemental processes will be obtained, and it is possible to define proper prevention measures that act on those processes.

© The Author(s), under exclusive license to Springer Nature Switzerland AG 2020
N. Fernandez-Anez et al., *Explosion Risk of Solid Biofuels*,
SpringerBriefs in Energy, https://doi.org/10.1007/978-3-030-43933-0_9

There are two types of prevention measures: organizational measures and technical measures. The first ones include employee's safety training, and safety manuals and procedures. The second ones are the physical application of the organizational measures. On the other hand, regarding the factor on which the measure acts, the prevention measures can be classified on.

9.1.1 Avoid Flammable Products

Regarding biofuels, it is impossible to avoid flammable products, as biofuel's nature is per se flammable. However, it might be possible to change secondary products used in biofuel's industry, in order to use non-flammable products or, at least, less flammable.

9.1.2 Reduce the Concentration or the Production of Fine Particles

Sometimes it is possible to change some processes or equipment in order to avoid the formation of fine particles and dust. Greater granulometries or less dust concentration, reduce the possibility of explosive atmosphere formation. In some cases, the use of additives might reduce the dust generation. Adequate equipment maintenance may help to reduce dust formation.

9.1.3 Reduce Accumulation of Dust

Dust accumulation may lead to self-ignition or dust layer formation. In those cases where it is not possible to avoid fine particles generation, reduce dust accumulation becomes essential. Proper and regular cleaning, aspirating and dust collecting systems constitute effective measures to avoid dust deposition.

9.1.4 Avoid Dispersion

Regarding gas processes, air circulation systems avoids explosive atmospheres; air flow has the opposite effect when flammable dust is present as it produces the dispersion and generates a dust cloud. Effective cleaning and collecting systems are the best option when flammable dust is present; however, if the process requires the use of flammable gases, and the presence of flammable fine particles is not avoidable, a combined system of air flow and aspiration must be installed.

9.1.5 Act on Ignition Sources

Control ignition sources intends to avoid the inflammation of the explosive atmosphere once that is formed. It is necessary to control equipment temperatures, avoid sparks or isolate them, confine direct flames, etc. Direct heat must be replaced by indirect heat if possible, and the use of internal combustion engines must be reduced to areas where explosive atmospheres cannot be formed or using spark suppressors. It is also necessary to limit hot surfaces temperatures and detect incipient self-heating processes.

9.1.6 Inertization

If an atmosphere cannot propagate an explosion, it is said that the atmosphere is inert. It means that the oxygen content is below the threshold that allows the propagation. Inertization can be partial or total, and it is produced by adding an inert gas that substitutes oxygen (typically nitrogen or carbon dioxide).

Inertization does not need to be applied only to the atmosphere, but also to the dust. If the fuel dust is mixed with inert dust, the combustion is avoided. However, dust inertization requires great amounts of inert dust which makes this technique hard to implement.

9.2 Protection

Despite the prevention measures applied nowadays in industry, sometimes it is inevitable to meet the optimal conditions for the beginning of an explosion at a certain point of the installation [1]. Then, when the explosion occurs, the application of protection measures become fundamental to minimize the damages and the derived costs. Consequently, this section contains the main protection systems and methods most used in potentially explosive dust atmospheres.

As presented in the New Approach Directives discussed previously, the protection measures will be divided into two large groups, protection methods for equipment in explosive atmospheres and protection measures for installations in potential explosive atmospheres.

9.2.1 Equipment

1. Electrical equipment intended for use in potentially explosive atmospheres

The protection methods for electrical equipment are the set of constructive rules of electrical materials and equipment that ensure their use in explosive atmospheres. Depending on the nature of the explosive atmosphere where the equipment is intended to be installed there are two different classes. The equipment for use in dust atmospheres are classified as Class II.

To adapt the adequate protection to each equipment, it is essential to carry out a zone classification according to Directive 1999/92/EU. Table 9.1 shows the classification of zones for Class II, based on the probability of explosive dust atmosphere appearance.

This class is regulated by the standards EN/IEC 60079-14 and EN/IEC 60079-17. It establishes the next steps to apply the proper protection measures, starting with the characterization of the substances, the sites, and finally the selection of the equipment as presented in Table 9.2.

In addition to the category of the equipment, the protection modes are divided into three large groups depending on the method used to avoid the explosion. The relation between the methods and the protection modes are described in Table 9.3.

The modes presented in Table 9.3 can be applied to atmospheres of gases, vapors, mists and dusts, but the objective of this chapter is to describe the protection modes regarding equipment intended to use in explosive atmospheres of dust hence Table 9.4 resume the specific methods for dust substances.

Table 9.1 Areas classification class II (Directive 1999/92/EC)

ATEX presence Probability	Very high	High and normal	Low
Dust and fiber	Zone 20	Zone 21	Zone 22

Table 9.2 Equipment classification depending on the zone classification

Equipment category	Zone
Category 1	20, 21 y 22
Category 2	21 y 22
Category 3	22

Table 9.3 Relation between the protection method and the protection modes

Method	Modes
Explosion confinement	d
Separation of the explosive atmosphere and the energetic source	P, m, q, nR, nC
Reduce energy or prevent sparks or arcs	E, ia, ib, nA, nC

Table 9.4 Type of marking depending on the protection mode

Type of protection	Marking
Intrinsic safety	Ex iD
Encapsulation	Ex mD (maD, mbD)
Dust atmospheres	Ex tD

Fig. 9.1 Simplified scheme of the intrinsically safety method

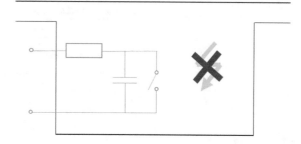

a. Intrinsically safety "iD":

The intrinsically safety is a protection mode largely used in industry. It is a measure taken directly in the electrical circuit ensuring that no spark, electric arc or thermal effect produced in normal operating conditions, as well as in specific fault conditions, can produce the inflammation of an explosive atmosphere. The industrial accidents caused by sparks are quite common [2] (Fig. 9.1).

Based on the specific test conditions the normative stablish two categories:

- Circuits "iaD": these circuits are not capable to produce an inflammation when K is:

    K = 1.5 Normal operation or one operational failure
    K = 1 Two simultaneous operational failures

- Circuits "ibD": these circuits are not capable to produce an inflammation when K is:

    K = 1.5 Normal operation or one operational failure
    K = 1 One operational failure with all contacts protected to avoid the appearance of sparks and may be exposed to an atmosphere explosive

where K is the safety coefficient.

Electric arcs, sparks, or reaching the ignition temperature are the possible sources of ignition in this case. Table 9.5 shows the categories in which the equipment is divided according to their ignition energy and their representative gases.

Due to the low energies, this mode will only be applied to small signal electrical and electronic circuits with small currents, voltages and powers.

The equipment heating is also considered. The temperature class of the equipment must be determined experimentally. This classification is carried out based on the maximum surface temperature of the equipment. This will always be the

Table 9.5 Equipment categories depending on the ignition energy

Group I (mine methane)	280 iJ
Group IIA	250 iJ
Group IIB	96 iJ
Group IIC	20 iJ

Table 9.6 Temperature classes of electric equipment

Temperature Class	Surface T_{max} (°C)	Inflamation T (°C)
T1	450	>450
T2	300	>300
T3	200	>200
T4	135	>135
T5	100	>100
T6	85	>85

highest temperature reached in the equipment service, considering the most unfavorable conditions, susceptible to produce the inflammation of the dust explosive atmosphere. Class II equipment shall be classified and preferably marked with the temperature class and, if necessary, limited to the specific combustible dusts for which the equipment is designed. The temperature classes are shown in Table 9.6.

A distinction must be made between "intrinsic safety material" and "associated electrical material." The first one has all the circuits under intrinsic protection, then it can be used completely inside the explosive atmosphere. The second one only has a part of the circuits under the intrinsic safety mode.

b. Dust atmosphere "tD":

It consists of confining the electrical material in an enclosure containing material in a powdery state. This powdery atmosphere ensures that no electric arc or heating that occurs inside can cause the explosion. Its use is not widespread, almost always limited to transformers, capacitors and electronic equipment as a complement to other protection modes. Its reparation is simple and low price (Fig. 9.2).

Fig. 9.2 Simplified scheme of the dust atmosphere method

Fig. 9.3 Simplified scheme
of the encapsulated method

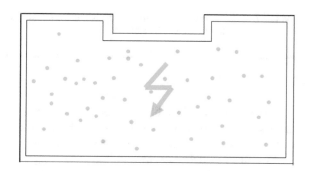

c. Encapsulated:

It consists of applying resin in those parts that can ignite an atmosphere by sparks so these parts are embedded. This protection method is especially useful for switchgear and small-sized equipment such as relays, sensors, capacitors, reactance, etc. In this case, it is not possible to carry out repairs or maintenance (Fig. 9.3).

d. General requirements:

Those type of equipment with protection methods must also comply with the general requirements set forth in the regulations [3]. These requirements are divided into common rules applicable to all electrical equipment and specific rules depending on the type of electrical equipment.

2. Non-electrical equipment intended for use in potentially explosive

Generally, non-electrical equipment is those that are capable of conduct their function without the use of electrical energy. The ignition sources associated with this type of equipment may be those listed below.

- Hot surfaces
- Exposure to the air
- Hot gases/liquids
- Mechanical sparks
- Adiabatic compression
- Pressure waves
- Exothermically chemical reaction
- Dust self-ignition
- Electric arches
- Static electrical discharges.

Therefore, protection measures shall be considered and applied in the following order:

1. Avoid the ignition sources.
2. Ensure that sources of ignition do not become effective.
3. Ensure explosive atmospheres do not reach sources of ignition.
4. Contain the explosion and prevent the spread of the flame.

Table 9.7 Relation between the protection method, the marking and the category for non-electrical equipment intended to use in dust explosive atmospheres

Protection method	Marking	Category
Constructional safety	c	M2, 1, 2, 3

In the case of combustible dust derived from the use and processing of biomass, there is only one protection method applicable as exposed in Table 9.7. This protection method consists of ensuring the absence of sources of ignition.

As well as electrical equipment, the general requirements are established in the regulations [3]. They are classified in common application requirements for all non-electrical equipment and specific rules depending on the type of non-electrical equipment. The most outstanding general requirements will be the determination of the maximum surface temperature, the control of dust deposits between moving and fixed parts of the equipment and the control of information on all potential sources of ignition. For the specific case of dust processing, the ignition risk due to the accumulation of dust from the process cannot be avoided, then the specific protection measures included in the EN 1127 [4, 5] series of standards must be taken.

Non-electrical equipment is classified in the next categories:

- Group I: equipment intended to use for mine Works or external facilities where explosives atmospheres can occur.

    - Category M1: very high level of protection.
    - Category M2: high level of protection.

- Group II: equipment intended to use in other sites with the risk of explosive atmospheres formation.

    - Category 1: very high level of protection.
    - Category 2: high level of protection.
    - Category 3: Regular level of protection.

As stated in Table 9.7, the protection method applied for non-electrical equipment in potentially flammable dust atmospheres is construction safety.

- Constructional safety "c":

This method consists of applying constructive measures to achieve safety against mechanical ignition caused by hot surfaces, sparks and adiabatic compression generated, for example, by moving parts (Fig. 9.4).

This protection method defines the equipment construction according to IP protection rating of enclosure protection established in EN 60523/A2:1999. Table 9.8 summarizes these protection rating for dust explosive atmospheres.

Fig. 9.4 Simplified scheme
of the constructional safety
method

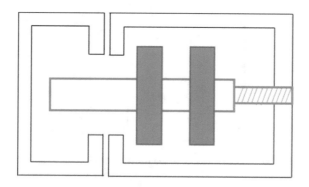

Table 9.8 IP protection depending on the equipment location

Equipment location	Risk	Protection rating
Non-flammable dust	Dust might enter the equipment	IP2X
Dust explosive atmospheres	Dust generate an ignition source or fire	IP6X
	Dust might enter the equipment but do not generate an explosive atmosphere	No protection

9.2.2   Installations

1.   Constructive protection systems

1.1.   Isolation or confinement of the explosion

Normally, when an explosion begins, the primary explosion can be propagated to other potentially explosive atmospheres. This fact is extremely dangerous since the pressure generated by the initial explosion suspends the dust deposited in nearby places, increasing the virulence of the explosion due to its rapid combustion [6]. By separating zones, compartmentalization or isolation, it is possible to confine the primary explosion.

Protective measures should detect, extinguish or block fire at an early stage, as explosions generally spread by flame and not by pressure waves. If the flame front meets other explosive atmospheres, the mixture will start immediately and explode. The result will be an increase in the combustion reaction and, therefore, in the explosion overpressure.

Fundamentally these protection systems consist of separating or isolating different areas of the installation as shown in Fig. 9.5. Enough minimum distance must be ensured between the recoupling that will separate the different areas of the process.

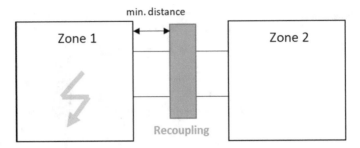

Fig. 9.5 General scheme of the isolation method for installation

The main protection systems are described below.

1.1.1. *Zone or equipment separation*

The separation of areas and equipment is one of the most basic, but most effective, protection measures in terms of dust explosions. By separating those equipment or installations, the primary explosion is limited, and its propagation is prevented. A clear example would be to locate the filters and cyclones outside the main installation. Equipment insulation must include the separation of possible conduits and connecting pipes so that the explosion cannot be propagated. The main equipment intended for isolate equipment or installations are listed below.

1.1.2. *Rotary closures and worm screw*

Firstly, the valves or rotary closures prevent the transmission of the flame and the pressure, from one side to the other of the closure (Fig. 9.6). Its construction must be carried out as listed below:

- The rotor fins must be metal and have a minimum thickness of 3 mm.
- The rotor construction must be robust enough to prevent it from moving in the radial or axial direction when an explosion occurs.
- The housing of the enclosure must be resistant to the overpressure that is generated in the explosion.
- The gap must be closed automatically in case of an explosion to avoid a secondary explosion as a result of the transport of a mass of incandescent material.

Secondly, worm screws prevent the spread of an explosion if the central part of a screw feeder is removed. In this way, the product mass acts as a closure or stopper that strangles the passage of any dynamic effect (Fig. 9.7).

1.1.3. *Devices to interrupt explosions*

Passive Devices

(a) Discharge chimney of the explosion or diverter:

This explosion decompression nozzle deflects the 180° of the air flow allowing the flame front and the explosion to be vented to the outside, preventing its propagation in the line of pipes and ducts (Fig. 9.8).

Fig. 9.6 Simplified scheme of a rotative closure

Fig. 9.7 Simplified scheme of worm screws

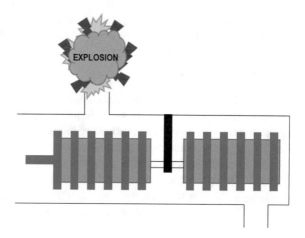

This protection system is mainly used in dust extraction lines to prevent the recoil of dust towards the entrance of the equipment. This method should be avoided if the powder is abrasive. In addition, this type of device only prevents the propagation of the "upstream" dust, then, if the explosion takes place at one of the dust exhalations points, the decompression nozzle will not prevent the propagation of the "downstream" explosion towards the filter. To avoid this situation, it is possible to add a quick slide valve or an additional barrier.

Fig. 9.8 Scheme of a
common discharge chimney
of explosion

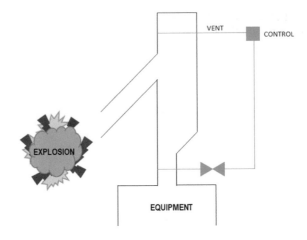

(b)   Detonation barrier:

There are two types, barriers against detonation and barriers against deflagration. The last one is installed under atmospheric conditions and at the end of the pipe since it is the point with the lowest working pressure and therefore with the lowest risk of fire in the protection against deflagration. On the other hand, detonation barriers will only be effective in short pipe sections. These types of barriers are adequate when the required response time is high (Fig. 9.9).

(d)   Ventex valve:

These valves are especially effective in the case of dust removal and grinding systems, very common systems in the biomass industry. This type of protection measures is applicable only for low concentrations of flammable dust and dust classified as St2. For St3 powders, it must be confirmed with the manufacturer (Fig. 9.10).

The minimum and maximum distances for installation are shown in Table 9.9. These distances depend on the explosion pressure (P_{max} or P_{red}) in the protected equipment, the nominal diameter of the pipe where it is installed, DIN, and the situation in which the combustible dust that is present is found. Therefore, the valve

Fig. 9.9 Simplified scheme
of detonation barriers

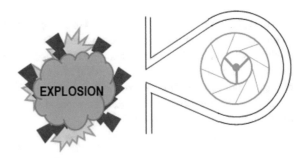

Fig. 9.10 Simplified scheme
of a Ventex valve

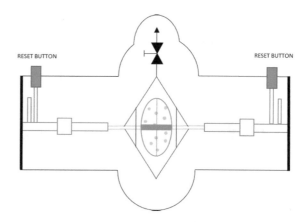

RESET BUTTON   RESET BUTTON

Table 9.9 Minimum and
maximum distance of the
valve location

Combustible dust		
DIN (mm)	D_{min} (m)	D_{max} (m)
100	5	12.5
200–700	5	12.5

should be positioned as indicated in Table 9.9. If this is not possible, it must be calculated based on the P_{max} of the protected equipment.

When the protected equipment is cartridges or sleeve filters, the effectiveness of the valve decreases as these systems positively affect the explosion. Normally, in these cases the valve can be installed at 0 m, that is, directly after the filters, since in case there is an explosion in the filter, the flame will ensure that the explosion pressure will reach the valve and will close it. In any case, the valve will be controlled by its corresponding control unit.

Active Devices

These devices act on the protection system by means of explosion sensors, usually flame or pressure sensors. The most common systems are explained below.

a. Active Ventex valve:

It is used in cases of low overpressure. They consist of an auxiliary gas flow control sensor.

b. Mechanical slide or quick fastening valve:

This valve is specially designed to prevent the spread of the flame front through the process ducts of the plants. Typical systems for placement are grinding, mixing, drying and spraying processes (Fig. 9.11).

c. Butterfly valve:

Its use is limited to equipment that has controls for the primary explosion such as venting or suppression of the explosion. The butterfly valve is then actuated by a

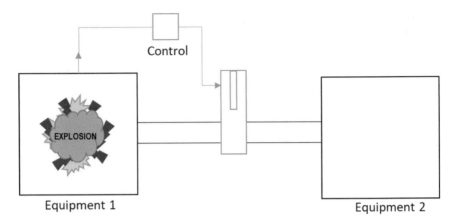

Equipment 1   Equipment 2

Fig. 9.11 Control system to avoid the propagation of the explosion using a mechanical slide valve

Fig. 9.12 Butterfly valve
functioning scheme

primary detection device thus stopping the pressure wave and the flame front that
arrive from the equipment (Fig. 9.12).

d.   Automatic termination system in pipelines:

By injecting a downstream extinguishing agent, the flame front is extinguished in the
inert zone in the pipes. The complete system consists of a detector (usually pressure
or optics detectors), a control unit and a bottle of extinguishing agent. The volume
of the bottle must be sized according to the need of the process (Fig. 9.13).

e.   Detection and extinction of sparks. Infrared technique:

This method focuses on the detection of airborne sparks or incandescent points in
pipes that connect points of the installation. This method is especially effective in
preventing the aspiration of sparks that can cause fires in filters. This system consists
of an optical detector, a controller and an extinguishing system, usually of water
spray.

Fig. 9.13 Control system
scheme for pipelines

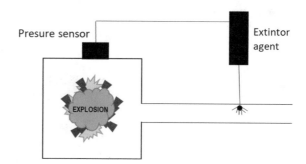

9.2.3   Explosion Suppression

The explosion suppression consists of extinguishing the incipient flame in the formation phase, to avoid its propagation. Consequently, an extinguishing agent is normally
discharged that prevents the overpressure that would be reached in a closed container
without suppression.

To detect the explosion in early stages, it is more appropriated to use pressure
detectors, although optical flame detectors can also be used. The response time must
be as fast as possible, always following the limits of the normal operation of the
equipment. For pressure detectors, the fundamental parameters will be P_{stat} which
will be the normal operating pressure of the detector and Pred, which will be the
pressure reached in the enclosure after the explosion.

Explosion suppression protections can be designed during the engineering project
stage or when the installation is already underway. In the first case, the best option is
to install standard suppression systems that are placed on the market and adapt it to
the specific requirements of the installation. The probability of failure will be very
small. In the second case, the most frequent in the industry, the installation is already
in operation, so the protection system must be designed in reverse. Its design will
depend on the geometry and mechanical characteristics of the installation, as well as
the specific conditions of each process.

When installing in the industry, its applicability limits must be considered. This
protection system is not suitable if the maximum rate of dust pressure increases
is very high since the system will not be fast enough to extinguish the explosion
before it spreads. In general, the higher the K_{max} of the powder, the more difficult
the suppression. The limit value of $K_{máx}$ for its application will be 300.

2.   **Devices and systems of the explosion venting**
2.1 **Facilities oversizing**

This method of protection consists in allowing the development of the explosion in
a safe and controlled environment. Its dimensioning is essentially divided into two
methods, the pressure resistant design and the pressure shock resistant design. The
first applies to vessels capable of withstanding the maximum permissible pressure,

Fig. 9.14 Simplified scheme
of the venting method

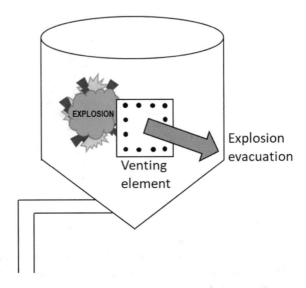

MPP, for long periods of time. The second for the same MPP allows deformation in
the container, so its construction is lighter than the previous one.

2.2 Explosion venting or decomposition

The vent is based on installing openings in the enclosure or container to be protected
so that, in case of an explosion, the pressure inside does not exceed the preset safety
limit value (Fig. 9.14).

Through a predetermined area (venting area), the pressure of the explosion is
released when a pressure level Pset is reached in the enclosure, resulting in the
decompression of the explosion, which leaves a residual pressure or pressure reduction Pred, which must always be less than the maximum pressure bearable by the
enclosure. This allows the flame, the pressure wave and the mixture to be released
without burning to a safe area. This implies that the explosion occurs, then, in the
case of combustible dusts, cleaning and confinement of the powder is essential since
this dust can lead to a secondary explosion. The design parameter of this type of
protection is the maximum pressure, which can usually be around 0.3 bar.

Venting is an economic protection method, but it should never be used in case of
toxic dusts. In addition, the vents must be closed during normal operations, preventing
the combustible dust handled in the process from leaving or that external agents, such
as moisture, can affect the dust inside.

For the design of the venting protection, two parameters must be considered,
maximum explosion pressure (MEP) and maximum rate of pressure rise. For a given
combustible dust, MEP and the maximum rate of pressure rise depend on several factors, such as dust concentration, particle size or cloud turbulence. MEP is practically
independent of the size of the recite where the pressure develops, while maximum
rate of pressure rise depends heavily on such volume; The "cubic law" establishes

the relationship that links these variables with the powder constant, K_{max}. For its correct design, the calculation of the venting area must be carried out using equations defined for each situation, including the following situations:

a. Pressure venting of equipment, including silos.
b. Venting of building envelopes.
c. Pipelines vent.
d. Explosion pressure vent of interconnected equipment with pipes.
e. Flame propagation.
f. Pressure of the propagation.
g. Effect of vent ducts on reducing the explosion overpressure.
h. Withdrawal force.

Depending on the elements used for the design of the venting equipment, these can be classified into the following groups.

1. With reusable items: They have the greater facility to define with precision the action pressure required for the opening. They are divided into the following subgroups:

    a. Devices that automatically close again
    b. Devices that need to manually replenish their explosion containment and pressure-sensitive elements.

2. With non-reusable items: Its main advantages are its low relative cost, good closure that they provide in their normal state, have a very low weight and provide fast-acting. There are the following types:

    a. Devices with a deformable rod
    b. Disposable panels
    c. Panels fixed at the edges
    d. Disks and panels with auxiliary activation.

3. Venting devices without flames: This equipment has special measures that allow to be venting the explosion being able to cause the extinguishment of the flame, avoiding the explosion propagation to its surroundings. They are used to allow the venting of the explosion in situations where a certain risk of fire after the venting is not acceptable, for example, in equipment located inside a plant and whose venting cannot be channeled outside through ducts.

References

1. Yuan Z, Khakzad N, Khan F, Amyotte P (2015) Dust explosions: a threat to the process industries. Process Saf Environ Prot 98:57–71
2. Randeberg E, Eckhoff RK (2006) Initiation of dust explosions by electric spark discharges triggered by the explosive dust cloud itself. J Loss Prev Process Ind 19(2–3):154–160
3. EN, EN-60079-0:2013 (2013) Explosive atmospheres. Part 0: Equipment. General requirements

4. AEN/CTN 163, UNE-EN 1127-1:2012 (2012) Explosive atmospheres-explosion prevention and protection-Part 1: basic concepts and methodology
5. AEN/CTN 163, UNE-EN 1127-2:2014 (2015) Atmósferas explosivas. Prevención y protección contra la explosión. Parte 2: Conceptos básicos y metodología para minería. (Ratificada por AENOR en Abril de 2015). Explosive atmospheres-Explosion prevention and protection-Part 2: Ba
6. Yuan Z, Khakzad N, Khan F, Amyotte P (2016) Domino effect analysis of dust explosions using Bayesian networks. Process Saf Environ Prot 100:108–116